Functional Analysis Tools for Practical Use in Sciences and Engineering

Carlos A. de Moura

Functional Analysis Tools for Practical Use in Sciences and Engineering

Springer

Carlos A. de Moura
IME – Instituto de Matemática e Estatística
Rio de Janeiro State University - UERJ
Rio de Janeiro, RJ, Brazil

This work was supported by FAPERJ - Fundação Carlos Chagas Filho de Amparo à Pesquisa do Estado do Rio de Janeiro E-26/202.500/2019

ISBN 978-3-031-10600-2 ISBN 978-3-031-10598-2 (eBook)
https://doi.org/10.1007/978-3-031-10598-2

Mathematics Subject Classification: 40A30, 46-XX, 46-01, 46B40, 46B50, 46F12, 46Nxx, 46N40

Translation from the Portuguese language edition: "Análise Funcional para Aplicações: Posologia" by Carlos A. de Moura, © Carlos A. de Moura 2002. Published by Ciência Moderna. All Rights Reserved.

This Springer imprint is published by the registered company Springer Nature Switzerland AG
The registered company address is: Gewerbestrasse 11, 6330 Cham, Switzerland

SANDRA,
my love cluster point,
how can a being to be
just as you are:
firm – as theorems
transparent – as an algorithm
beautiful – as a poem ...

To my father, JOÃO
Seu JOCA,
a solemn truck driver,
who has so much learned
from bumpy roads,
and from so many trips
he has fought through his life –
also a road
to continuous teaching ...

Preface (to the Portuguese Version)

Mathematical topics labelled as **Functional Analysis** – concepts, definitions, results – compose a deep example of a mathematical structure aimed at applications – problems originated from Physics, from Engineering Sciences, even from other areas within Mathematics. That is also the purpose carried by this book: to bring familiarity to the reader who has been dwelling on topics where these tools are needed, but definitely who does not wish to develop research on the associated theory. Instead, as long as safety and precision are strongly required, look forward to be kept safe of fallacious slips and misconceptions that a more shallow contact often leads to.

The text emphasizes motivation, justification for the choices made, the right way to employ discussed techniques, but in a large number of spots, proofs are not shown. In such cases, the preferred author's references are pointed. Complete proofs – or sometimes only their sketch – get exposed whenever fared their knowledge and familiarity as imperative for a safe use of the results they claim. Or else when they indicate a technique which itself will bring its presence in any of these very applications.

A main thread may be found throughout this exposition: to link the ideas inside the **completion of a metric space** to those on the **continuous extension of operators**. That is the tool that lets us, for example, to introduce in a quick way Lebesgue integral, Sobolev spaces, and that leads to the almost ubiquitous regularization technique, or the *mollifiers*, in K.O. Friedrichs' favorite terminology.

This book is the third version of [22], written for a short time school and which, due to the lack in Brazil, at that time, of texts with this approach, was later used in different academic environments, until a second version [23] was published. This one took on the same road, and with the suggestions of many colleagues that have used it, I faced the adventure of the present write-up, more careful and with some additions. The previous versions fail to include applications because these were part of other mini-courses presented at the schools – this and those other ones have been written for. Now this version includes additional applications, all related to Numerical Analysis – finite elements, stability for numerical solutions for differential equations. Besides, the presented results about Lebesgue integral have been put together and are more complete. The Fourier and Laplace transforms, together with the tempered distributions, are now included.

The sharpness of the mathematical remarks as well as suggestions about writing language and exposition view, thanks to the previous careful reading carried on by Professor Dinamérico Pombo – a very particular friend of mine, then at UFF Mathematics Institute, was of out of sight value.

To Professor David Isaacson, from Rensselaer Polytechnic Institute, I owe a dialogue with such a convincing power, which I always brought it back to my memory when the giving up option was pulling me out.

A large portion of this book was written when the author was visiting the Mathematics and Computation Institutes of the Fluminense Federal University – UFF. I would like to express my thankful words for their support.

The book was finished and published while I was a visitor at the cozy environment from IME-UERJ, the Mathematics and Statistics Institute of the Rio de Janeiro State University. The valuable support must also be acknowledged.

Conventions

Our notation follows the mathematical text standards, either theoretical or from applications; exceptions are described whenever introduced, or in the paragraphs that follow. We make use of the following conventions, all usual.

$\mathbb{N}$: the positive integers
$\mathbb{Z}$: the integers
$\mathbb{Q}$: the rationals
$\mathbb{R}$: the real line
$\mathbb{R}^N$: the Euclidean space with N dimensions
$\mathbb{C}$: the complex plane

For two sets A and B, we denote by $A\backslash B$ the third combining set – counting $A \cup B$ and $A \cap B$ – of elements from A that do not belong to B.

Given the reals a, b, with $-\infty < a \leq b < \infty$, we denote by $[a, b]$ the closed interval $\{x \in \mathbb{R}; a \leq x \leq b\}$, while for the open interval $\{x \in \mathbb{R}; a < x < b\}$ we shall use either $]a, b[$ or (a, b), with analogous convention for the semi-open intervals.

Sequences will be denoted by $\{x_n\}_{n\in\mathbb{N}}$, $(x_n)_{n\in\mathbb{N}}$ or, in a simpler way, by $\{x_n\}$ or (x_n), even using upper indices. Not much strictness either with the notation for functions: $f(x)$, $f(\cdot)$, or f.

The symbol $:=$ in a given expression means that the "right-hand side" defines whatever occupies the "left-hand side." As regards to the symbol •, to point to the end of some topic, its use was quite stingy, it only shows up if its absence would bring doubts to the reader. Observe that a relative of this is used to indicate which is the main argument in some expressions, like $f(\cdot)$, $\|\cdot\|$ or analogous ones.

To finish, it is wise not to take the notation for granted, as long as we have allowed it being "*una donna mobile*". Indeed, it may exhibit changes from chapter to chapter, whenever convenient, as for example in:

$$\|f\|_{L^2} = \|f\|_0 = \|f\|_2 = \|f\|_{0,2} = \{\int f^2\}^{1/2},$$

$$\|f\|_{H^1} = \|f\|_1 = \|f\|_{1,2} = \{\int f^2 + \int (f')^2\}^{1/2}.$$

Itinerary

With Chap. 1 we look to motivate and, to some extent, to justify the developed theory being announced. Chapter 2 introduces or recalls definitions and notation for abstract spaces as well as for specific function spaces. This endeavor requires Lebesgue integral, whose development for the real line through an almost *geodesic* way, we discuss. And try to leave clear the changes needed to apply the results and proofs for $\mathbb{R}^N$.

The aim in Chap. 3 is to discuss some properties of dual spaces, which are firmly glued to those particular spaces presented in the previous chapter; in particular, the classical – and indispensable – identifications that allow to operate with them. They turn out to be tools frequently used in Chap. 4, where ideal, or generalized, functions are introduced. They allow us to refer with the needed rigor to, for example, the *Dirac delta*. Thereby, we also present the construction for Sobolev spaces besides introducing the Fourier and Laplace transforms.

The so-called Linear Functional Analysis *basic triplet* is the core of Chap. 5, some applications being discussed therein. Chapter 6 introduces the concept of compactness, fundamental for the construction of approximation sequences, as long as it allows to guarantee convergence, in many contexts.

Up to this point, this text reading is assumed to be made sequentially. Now, Chap. 7[1] uses only concepts introduced on Chap. 2, while the last one depends on concepts from Chaps. 2 and 3. Chapter 7 generalizes the concept of derivative for functions defined on normed spaces. Such a generalization allows to use, on these spaces, approximation algorithms, like Newton-Raphson's, besides several optimization methods.

In the last chapter, one finds basic results about Hilbert spaces that turn out to be of heavy importance to numerical approximations as well as to applications on Physics, e.g., Quantum Mechanics. Within this framework, spectral theory is discussed only for compact operators, so that we suggest for more general browsing

[1] Portuguese version had it as the last chapter. We allowed ourselves here to adapt the translation data as this version has exchanged it with the chapter previously taken as the seventh.

[6, 72]. These ones are also reference for the semi-group theory which, despite its importance in several applications, see [36, 49], we lacked to bring into discussion.

Rio de Janeiro, Brazil
October 24th 2002

Carlos A. de Moura

P.S.: contacts would be welcome at
demoura@ime.uerj.br

Preface to the English Version

Taken the decision to write myself this translation of [25], I thought at once that, for this task, no sense to apply the old cliché "*traduttore = traditore*". But quickly I realized that it would be hard to restrain from scattered minor changes, besides two sensible ones: the addition of the last section to the last chapter as well as, in the first chapter, the 5th encounter to the first section.

Despite a lot of time gone since the previous version, I fail to find much to add to the above translated preface. Nevertheless, I ought to recognize the four corners of a square. Each one lodges an unforgettable help well.

The first is occupied by my wife Sandra. Impossible to believe how she acted to bring joy to these pandemic days (and that is as she always acts!). Without her push, no end for this task...

A long-time friend – life has turned him my brother –, Jerry A. Goldstein, made me so happy to have him installed at the second corner, testifying the decades we have shared. He definitely convinced me it would be worth to put hours and thoughts on this version.

The support from UERJ, the university I work for since this millennium started, particularly from IME, its math institute, with friends and friendly colleagues, fills up the next corner.

The last one settles all *sine-qua-non* scientific/financial help from FAPERJ – the Carlos Chagas Filho Rio de Janeiro State Research Support Foundation – which has provided a grant that became essential. It is a pleasure to acknowledge that support filed by:

E-26/202.500/2019, Program for Visiting Professor PV-2019.

Rio de Janeiro, Brazil
September 8th, 2020
demoura@ime.uerj.br

Carlos A. de Moura

Contents

Notation

$\ell_0^\infty,\ \ell^\infty,\ \ell^2$
$c,\ c_0$
$C^k[a,b],\ C^\infty[a,b],\ C_0^\infty(I\!R)$
$\mathcal{S}(I\!R)$
$\|x\|_p\ \ (p=1,2,\infty)$
$\|f\|_p,\ \|f\|_{r,p}\ (p=1,2,\infty)$
$\delta_{t_0}(f)$
$B[v_0;r],\ B(v_0;r)$
$\ell^p,\ \|x\|_p\ \ (p\in[1,\infty))$
$\text{dist}\,(A,B)$
$\mathcal{L}(X,Y),\ \mathcal{L}(X)$
$ker(T)$
$Im(T)$
$\mathcal{F}(f)(t)$
$\mathcal{L}[a,b],\ \mathcal{R}[a,b]$
Ψ_A

$L^p(I\!R)$
$L^\infty(I\!R)$
$V',\ V^*$
$J,\ J_v$
D^ι
$H_0^k(\Omega),\ H^k(\Omega)$
$\mathcal{D}(\Omega)$
$\mathcal{D}'(\Omega)$
$L^1_{\text{loc}}(\Omega)$
$L^p\ (0,T;B)$
$\mathbf{o}(\ ||h||\ )$
supp
$\mathbb{I}$
$\Re(p)\ +\ i\Im(p)$
$\mathcal{L}[f]$
$[w]$
$[A]$

Remark: Some symbols employed in a single section fail to be listed herein.

Chapter 1
Road Map

Unfortunately, among scientific literature texts, we can find many reports of applications, as well as of theoretical results, fully based on a mathematical formalism guided by the three (**false**) principles hereby quoted.

- *Principle of universal permutability* – Whenever evaluating any combination of integrals, series, derivatives, or limits, the adopted order is irrelevant. The reached results are an invariant, no matter the chosen track.
- *Principle of analogy between discrete and continuous indices* – Properties held by finite sums must remain valid for corresponding (or associated) integrals.
- *Principle of unrestricted convergence* – Whatever sequence, series, or improper integral one deduces within a theoretical development, it ought to be convergent, unless a mistake has been introduced in the construction of the mathematical model thereby employed.[1]

1.1 Some Encounters

Mathematical theory evolution is led by two mutually competing types of demand: the first ones are internal and grow up due to all kinds of questionings. They are created within the structure of its own concepts, definitions, and conclusions; in short, they look for pure mathematical results. The other ones, external, are born inside different scientific or technological areas that get hold of some mathematical environment for their theoretical advancement, or else to simplify their own procedures. And it turns out that, quite frequently, as fruit of this latter interaction, unexpected forward laps have been observed. These successful results have arisen from the need to fill up the gap of still unavailable – but sometimes wrongly dreamed of – mathematical concepts or elements.

[1] The text above is a free translation of [30], page *xi*.

C. A. de Moura, *Functional Analysis Tools for Practical Use in Sciences and Engineering*, https://doi.org/10.1007/978-3-031-10598-2_1

But it should also be underlined that it is quite frequent, despite rather dangerous, the ill use of correct results, as well as the calling of ill-stated claims – like those ironically quoted in the opening of this chapter. It is impossible to preview the associated consequences, but anyway, the conclusion's logic will always be wrong. Besides, being the mathematical tools supported by an increasingly powerful computer systems, it becomes practically impossible to evaluate experimentally or computationally such deduced conclusions.

The examples that follow aim to illustrate some topics to be discussed afterwards, or either to motivate such discussions. At the same time, they light a warning about the serious risks brought in by the abovementioned *false principles*.

Encounter 1: No Matter Which Order? A fact taught from Calculus courses: convergence of the so-called **alternate harmonic series**

$$h := \sum_{\imath=1}^{\infty} \frac{(-1)^{\imath}}{\imath}. \tag{1.1}$$

Suppose somebody, after stumbling on the series

$$-1 + 1/2 + 1/4 - 1/3 + 1/6 + 1/8 + 1/10 + 1/12 - 1/5 + 1/14 + 1/16 + \ldots,$$

observes that all its terms are the same as those from the alternate harmonic series, just with a reordering. It may be accepted as rather natural to conjecture that this latter is also a convergent series and further that it gets the same sum. Then, a question pops up at once: for a sum of *infinite terms*, does commutativity and associativity keep holding, as it were a finite sum? Or, in more precise terms, given a permutation of the naturals, *i.e.*, a bijection from the set of natural numbers on itself

$$\tau : \mathbb{N} \to \mathbb{N},$$

is it necessarily valid for such rearrangement of (1.1), that

$$\sum_{\imath=1}^{\infty} \frac{(-1)^{\tau(\imath)}}{\tau(\imath)} = \sum_{\imath=1}^{\infty} \frac{(-1)^{\imath}}{\imath}?$$

A simple question with a surprising answer! Depending on the permutation τ, everything may show up: the new series may either diverge, or else it may converge, but towards no matter which value is chosen on the line (or even on the extended line). Let us describe this result in a more precise fashion.

For any real γ taken on the whole line, there exists a permutation of the natural numbers, $\tau = \tau_{\gamma}$, for which

$$\gamma = \sum_{\iota=1}^{\infty} \frac{(-1)^{\tau(\iota)}}{\tau(\iota)}.$$

It is worth mentioning that this result is valid not only for the alternate harmonic series but also for all **conditionally convergent series** of real numbers, cf. [61], Theorem 3.55, where the more general formulation we have first quoted is discussed. The proof therein presented exhibits a nice sample of a constructive reasoning within this framework.

This example deserves to be thought of as a herald for a warning: far from being a mere **addition** of numbers, an infinite series is indeed a **limit** process!

Encounter 2: To Mend a Torus? A torus in $\mathbb{R}^3$ may be identified with the whole plane $\mathbb{R}^2$ throughout quite a simple consideration: associate the points (t, y) in the unit square $\mathcal{Q} := [0, 1[\times [0, 1[$ to the points on the plane by identifying

$$\left.\begin{array}{l} t \quad \text{with } t+n,\ n \in \mathbb{Z} \\ y \quad \text{with } y+m,\ m \in \mathbb{Z} \end{array}\right].$$

An initial feeling may drive one to expect that this shrinking – a three-dimensional body associated to a two-dimensional space – is the most one may reach. It should not be counted upon, for example, that it is feasible to thread a piece of line all around the torus and completely cover it, so as to have any hole it carries, no matter how small, mended, hidden. But just be surprised, that is precisely what occurs, as described now.

We need to deal with the family of curves $\mathcal{C}_{\alpha,\beta}$ defined by

$$\mathcal{C}_{\alpha,\beta} := \{(t, y(t)),\ y := \alpha + \beta t;\ -\infty < t < +\infty, 0 \le \alpha < 1, \beta \in \mathbb{R}\},$$

on $\mathbb{R}^2$, taking hold of the just introduced identification. In other words, consider these curves image on the surface of the $\mathbb{R}^3$ torus and let us examine their behavior.

As long as $\beta \in \mathbb{Q}$, a strain-free verification tells that the corresponding curve $\mathcal{C}_{\alpha,\beta}$ **is** periodical on the torus. The same reasoning also implies that, whenever $\beta \notin \mathbb{Q}$, then $\mathcal{C}_{\alpha,\beta}$ **is not** periodical. Moreover, an additional effort will get, for every $\alpha \in [0, 1)$ and $\beta \notin \mathbb{Q}$, the following conclusion about the curve $\mathcal{C}_{\alpha,\beta}$. No matter which point on the torus is chosen, as well as which distance bound is required, there exists a point on this curve whose distance (on the torus) to the previously chosen point shows up to be smaller than the beforehand tolerated distance.

A quite important feature this example is sought to emphasize: small changes in one of the **input** parameters from a particular mathematical model may lead to extremely strong changes in the **output** data. In the just discussed case, we have moved from curves that carry a light *weight* on the torus to other ones that show themselves *everywhere* in this torus, or loosely said, that "fill up the whole torus."

It is also worth underlining that the **instability** presented here occurs within a very simple environment, we could even roughly describe it as linear.

Encounter 3: Functions, Distributions, ..., Surprises In[2] 1872 Weierstrass hugely shocked the mathematical community when he proved the existence of an *everywhere continuous* real function which fails to be differentiable on all of its domain, the whole line. An example of such a function may be defined, cf. [9, 70], as the limit of the **uniformly convergent** series of **infinitely differentiable** functions

$$f(x) := \sum_{\iota=0}^{\infty} \frac{\cos(3^\iota x)}{2^\iota}.$$

Later on, in the 1930s, while searching for the mathematical foundations of a model for the mechanics of atomic particles, Paul Dirac introduced a *function* subjected to quite special conditions, namely,

$$\left.\begin{array}{ll} \delta(x) = 0 & \forall x \neq 0 \\ \int_{\mathbb{R}} \delta(x) f(x) dx = f(0) & \forall f \in \mathcal{F} \end{array}\right],\tag{1.2}$$

where $\mathcal{F}$ denotes a conveniently *a priori* defined function set, cf. [29].

As long as in (1.2) the integrals shown are of Riemann or Lebesgue type, no *traditional* function may fulfill both conditions. Nevertheless, for a long period, physicists have (successfully) employed this mathematical being – considered queer – as it failed to own duly structured basis. And despite having the mathematicians stayed uncomfortable towards this structure, it remained being applied over and over, before a formalization was born. This way, it has even driven to theoretical results that later on bore experimental confirmation. This success leads to the spreading of the terminology Dirac *δ-function*, or simply *δ-function*.

Working independently, Laurent Schwartz [65] and other mathematicians have untangled the knot *δ-function* has brought in. The stroke was in some sense unexpected: just look at these objects as of a new kind, as generalizations for *bona fide* functions. Afterwards, they were baptized as distributions, or generalized functions.

As compared to functions, distributions show a behavior quite different with regard to the limit processes. For example, suppose a distribution T is a sum of a given series

$$T = \sum_{\iota} T_\iota.\tag{1.3}$$

It is possible (and correct) to get for T its derivative of any order – which always exists – just by term-wise differentiation of (1.3), or explicitly:

[2] Free version of [7], pp. 208, 216.

$$T^{[n]} = \sum_{\iota} T_{\iota}^{[n]}.$$

Definitely, a dramatic off-balance exists between the way functions and distributions perform within many frameworks. The behavior just pointed out must be compared with the one shown by Weierstrass function.

Encounter 4: Space-Filling Curves Consider a special type of simple curves on the plane, namely, the graph of a continuous real function whose domain is a bounded interval. A supposedly intuitive way to characterize such an environment (the set of continuous functions) is to require that each of their graphs may be drawn by keeping a pencil steadily on a sheet of paper, never taking the hands off the sheet until the graph is done.

This lousy definition reinforces the idea one naturally absorbs about the set of the (domain) points where a continuous function lacks to be differentiable. One is tempted to think of it as a discrete set, where two of its members, whenever "neighbors," are split by a whole interval. And throughout that interval, the considered function holds differentiability everywhere.

Weierstrass function, **W'f**, mentioned at the previous example, guarantees that some functions, despite continuous, may show themselves as roughly misbehaved. Certainly for **W'f**, pencil and paper graph drawing is far from an achievable task... Observe a sensible distinction between the building of **W'f** and of the torus-filling functions on Encounter 2. These latter ones were presented throughout an explicit construction, based on a finite sequence of steps. On the other hand, **W'f** was introduced with the mighty alternate of a uniformly convergent trigonometric function series.

In the sequel, an example is indicated of a space-filling plane curve, namely, a continuous mapping from a real bounded interval whose image fills out a plane square. Guided by our intuition, we ought to expect that a plane curve, being locally a finite sequence of not-too-deformed intervals, owns the look of the geometrical body we are used to call "curve." In order to reach a different dimension body – a square – we must be driven by an "infinite" procedure, climbing up to a limit process. And again the solution is brought from a uniformly convergent function series.

It is worthwhile to point out that the results hereby presented – as well as many similar ones – may be linked to sometimes unexpected applications. It is quite common to look at surprising examples – or counterexamples – as results *per se*, but in fact they belong to a more involved stream of reasoning and search. The one below, discovered by Hilbert in the late years of the nineteenth century, was recently called in to design an efficient algorithm for numerically solving partial differential equations of a special type, cf. [11, 26].

The graphs of the six first functions of a sequence of curves whose limit is dense in the plane unit square – informally referred as "which fills up the unit square" – may be seen, say, at the link which follows:

https://upload.wikimedia.org/wikipedia/commons/thumb/7/7c/Hilbert-curve_rounded-gradient-animated.gif/440px-Hilbert-curve_rounded-gradient-animated.gif

Any reader who has ever been exposed to fractal sets of ideas has certainly recognized their presence in the construction suggested by the abovementioned pictures.

An additional remark: a rather naïve analysis for the curve hereby discussed would not assign it to the hall of real surprising ones. Indeed, just take into account that any piece of cloth – which may be thought, in a rigor freeway, as a plane body – is woven with, no rigor again, a finite bunch of "one-dimensional" thin pieces of thread. But it ought to be pointed out, as a response, that neither the cloth is two-dimensional nor the thread has one dimension only: they can at most be taken as approximate models for these mathematical abstractions. That was the reason for having chosen above the expression "jumping to the limit" – or preferably "diving to infinity" – so as to underline this unusual theoretical track from one towards two-dimensional spaces while searching such an unexpected function.

Encounter 5: Rationals – The Gaps They Leave One of the pieces of surprise the set of rational numbers play for its observers is the existence of gaps when its elements are lined up. Indeed, given two of them, say $a, b, a < b$, no matter how close they are, we have an "infinite amount" of rationals inside the interval $[a, b]$. In spite of this, many constraints may be imposed that the rationals fail to fulfill. The Greeks already knew to be impossible for a rational λ to satisfy the condition $\lambda^2 = 2$. This number surely is the most well-known irrational, always quoted in order to confirm the existence of the abovementioned gaps. Thus, it is worth to indicate other simple examples of them, like all the ones in the countable set $\{\rho \in \mathbb{R} | 2^\rho = 2n + 1, n \in \mathbb{N}, n > 0\}$.

Let x_0 be a chosen rational number, $0 < x_0 \in \mathbb{Q}$, to which a sequence of rationals is associated, namely, $(x_n)_{n \in \mathbb{N}}$ defined as

$$x_{n+1} := \frac{1}{x_n} + \frac{x_n}{2}, n \geq 0.$$

Just call for a little help from the function $\phi(x) := 1/x + x/2$ together with its derivative: their analysis will indicate that distinct positive rationals x_0 for which $x_0^2 < 2$ lead to distinct sequences. Besides, for all of them, it may be seen that, as long as $n \geq 1$,

$$\left. \begin{array}{r} 0 < x_n \in \mathbb{Q}, x_n^2 > 2 \\ x_{n+1} < x_n \\ \text{if } x_n^2 > \beta \forall n, \text{then } \beta \leq 2 \end{array} \right] . \tag{1.4}$$

In short, we have constructed strictly decreasing sequences of positive rationals. Moreover: all of them are subjected to the bound described by the last relation in (1.4).

Now observe: if any of these sequences converges, its limit ζ ought to fulfill

$$\zeta = \frac{1}{\zeta} + \frac{\zeta}{2},$$

an identity which implies

$$\zeta^2 = 2. \tag{1.5}$$

Therefore, as long as this limit exists, it fails to be a **rational number**.

Based on the knowledge of that the sequences $(x_n)_{n\in\mathbb{N}}$ are decreasing and lower bounded, we can claim that any of them is a Cauchy sequence[3]. Further, it can be verified that, for an arbitrary pair of these sequences $(x_n)_{n\in\mathbb{N}}$ and $(\tilde{x}_n)_{n\in\mathbb{N}}$, the limit

$$\lim_{\iota\to\infty} |x_\iota - \tilde{x}_\iota| = 0 \tag{1.6}$$

must hold. The whole bunch of just constructed sequences $(x_n)_{n\in\mathbb{N}}$ may be **identified** among themselves, each one to whatever another be chosen. Besides, all of them own a strong reason to be also identified to the number ζ. Recall that such a number is known to be **absent** from the set of rational ones – as remarked, (1.5) holds for itself. Such identification ought to include all Cauchy sequences $(\tilde{x}_n)_{n\in\mathbb{N}}$ that fulfill (1.6) for at least one of the sequences $(x_n)_{n\in\mathbb{N}}$ we have constructed. As a consequence, this identification must hold for all sequences obtained from our receipt. We can think that all these sequences are pointing, waiving towards this gap whose existence in the set of rational numbers we have become aware of.

This is one of the tracks to formally introduce the real line: add to the set of rational numbers the set of irrational ones, *via* construction of a larger set where all Cauchy sequences do converge.

1.2 Feel Invited

The track followed throughout the construction described on Encounter 5 is adopted in quite a number of spots in this text, at different contexts. They are more general than the set of rational numbers, despite being all of them heirs of the strategy therein developed. It is a whole framework which allows to conclude such constructions in an elegant fashion and quite quickly. The everywhere present stick to help the reasoning is the **Principle of the Continuous Extension**, to be described in the chapter which follows.

The above examples have steadily exhibited the task of identifying distinct concepts or sets, making easier their understanding or operation. We could even

[3] The geometrical series is a handy tool for convincing one of this result.

state that maybe this is the most important aspect of mathematical research: to search, among different **structures**, the one which allows to assign the same label or **identification** to apparently distinct, unconnected properties, concepts, or sets. Moreover, we let ourselves to allow the term identification to include **approximation**, where interchange is carried out for elements that bear intrinsic differences.

Let us observe that distinct ways of measuring distances – the so-called metrics – are inherent to the approximation concept. Nevertheless, emphasis must be put on the fact that not only **quantitative** aspects of approximation should be weighted by the chosen metric: the model under exam may require that other fine aspects should also be preserved, a **qualitative** evaluation ought to be performed then. We have in mind several constraints like conservation or decay laws, or some profile restrictions (monotonicity, bounds, regularity).

To browse through a mathematical framework demands always scattered reasoning about **functions**, and these are precisely the elements to be studied, dissected, grouped, and approximated, in short, **analyzed**. Let us share then a stroll on the **Functional Analysis** alley.

Chapter 2
Basic Concepts

Let us open this chapter recalling some concepts from linear algebra. We count upon the reader familiarity with them, but the purpose is indeed to have notation and terminology once for all fixed. Besides, the accompanying examples serve not only to clarify the introduced definitions but rather to help the study of applications to be shown in due time. Another point: it must be taken into account that the **exercises** collection play the role of main actors inside the discussed theory. To think about this cast is a need. Even if some of them fail to be solved, deeply think about their formulation. They are an essential portion of the text, and thus they will be certainly employed later on.

2.1 Real Vector Spaces

Recall that a **real vector space** is an arbitrary set V for whose elements two operations are defined, just as we have for the **vectors** in the three-dimensional space – namely, an **addition** and a **product by a scalar**. These operations are supposed to fulfill the following conditions.

(i) Given two arbitrary elements v_1, v_2 in V, their **sum** is another uniquely defined element from V, denoted by v_1+v_2, and for which the expressions below always hold true.

  (*a*) $v_1 + v_2 = v_2 + v_1$;

  (*b*) $(v_1 + v_2) + v_3 = v_1 + (v_2 + v_3)$;

  (*c*) Besides, there exists a unique element which we call **zero** and denote by $\dot{0}$, such that, for all v in V, the identity which follows is true:

$$v + \dot{0} = v;$$

C. A. de Moura, *Functional Analysis Tools for Practical Use in Sciences and Engineering*, https://doi.org/10.1007/978-3-031-10598-2_2

(*d*) Further, given an arbitrary vector v in V, there exists another element, called **symmetrical** of v and denoted by $-v$, for which

$$v + -v = 0.$$

(*ii*) Given a real α and an element $v \in V$, associated to this pair is one (equally unique) element from V, the **product of α by** v, denoted as αv, in such a way that, being $\alpha_\iota \in \mathbb{R}$ and $v_\iota \in V$, arbitrarily chosen, we have:

(*a*) $\alpha(v_1 + v_2) = \alpha v_1 + \alpha v_2$; (*c*) $(\alpha_1\alpha_2)v = \alpha_1(\alpha_2 v)$;
(*b*) $(\alpha_1 + \alpha_2)v = \alpha_1 v + \alpha_2 v$; (*d*) $1v = v$.

It is possible to deduce from these properties that:

$$0v = 0,$$

as well as that

$$(-\alpha)v = -(\alpha v).$$

Observe that the symbols "$-$" as well as "0" have distinct meanings in the two sides of these expressions.
The identity

$$\alpha 0 = 0,$$

may also be proven, as well as uniqueness for the vector zero, a property which is equally true for the symmetric of an arbitrary vector v. Moreover:

$$\alpha v = 0 \Longleftrightarrow \alpha = 0 \text{ or } v = 0$$

holds, or in an equivalent expression,

$$\alpha v = \beta v \Longleftrightarrow \alpha = \beta \text{ or } v = 0.$$

Underlining: the elements in V are called **vectors**, the vector zero is the **null vector** in the space under study, and the operations thereby introduced are called **sum of vectors** and **product of a real number by a vector**.

Example 2.1 The set of all N-tuples of real numbers $x := (x_1, \ldots, x_N)$, $x_\iota \in \mathbb{R}$, gets the notation $\mathbb{R}^N$, provided the vector operations be defined component by component (as in the familiar cases of $N = 2, 3$).

Example 2.2 The previous example is generalized by considering several sets of sequences of real numbers

$$x := (x_1, x_2, \ldots) = (x_j)_{j \in \mathbb{N}},$$

with distinct constraints, but for all of them being the operations also defined component-wise:

$$x + y = (x_\iota)_{\iota \in \mathbb{N}} + (y_\iota)_{\iota \in \mathbb{N}} := (x_\iota + y_\iota)_{\iota \in \mathbb{N}},$$

$$\alpha x = \alpha (x_\iota)_{\iota \in \mathbb{N}} := (\alpha x_\iota)_{\iota \in \mathbb{N}}.$$

(a) ℓ_0^∞ := the set of **almost null** sequences, which means

$$x \in \ell_0^\infty \iff \left[\begin{array}{l} \text{there exists } N = N(x) \text{ for which} \\ x_\iota = 0 \text{ if } \iota > N. \end{array} \right.$$

(b) $\ell^\infty := \{x = (x_\iota)_{\iota \in \mathbb{N}};\ |x_\iota| \leq M = M(x)\}$, the set of **bounded** sequences. (Observe that to each sequence in ℓ^∞, a distinct bound is associated!)

(c) c := the set of **convergent** sequences.

(d) c_0 := the set of convergent sequences **with null limit**.

(e) ℓ^2 := the set of sequences of real numbers whose squares compose a convergent series; in other words, the sequences

$$x := (x_\iota)_{\iota \in \mathbb{N}} \text{for which} \sum_{\iota=1}^{\infty} |x_\iota|^2 < \infty.$$

In the list below, each (above defined) space is properly contained in the one which immediately follows itself:

$$\ell_0^\infty \subset \ell^2 \subset c_0 \subset c \subset \ell^\infty.$$

Example 2.3 Denote by $\mathcal{P}_N$ the set of all polynomials of degree $< N$, taking for two elements $p(x), q(x) \in \mathcal{P}_N$ the vector sum defined as the usual sum of two polynomials: for

$$p(x) := \sum_{j=0}^{N-1} a_j x^j, \ q(x) := \sum_{j=0}^{N-1} b_j x^j,$$

then

$$(p + q)(x) := p(x) + q(x) = \sum_{j=0}^{N-1} (a_j + b_j) x^j.$$

In the same track, for a real α,

$$(\alpha p)(x) := \alpha p(x) = \sum_{\jmath=0}^{N-1} \alpha a_\jmath x^\jmath.$$

In all examples of function spaces that follow, the operations of sum and product by a scalar will be taken as pointwise, or explicitly,

$$(f_1 + f_2)(x) := f_1(x) + f_2(x),$$

$$(\alpha f)(x) := \alpha[f(x)].$$

Example 2.4

(*a*) Consider the set of continuous functions defined on [0,1], usually denoted by $C^0[0, 1]$. In general we can as well take $C^k[0, 1]$, $k \geq 1$, the set of functions defined on [0,1] and whose derivatives up to order k exist and are continuous. It is usual to employ the notation

$$C^\infty[0, 1] := \bigcap_{k=0}^{\infty} C^k[0, 1].$$

The functions in such a space are called **infinitely differentiable**.

(*b*) Take now the infinitely differentiable functions defined on the whole real line and that vanish outside a bounded interval – this interval changes with each considered function. Such a set is denoted by $C_0^\infty(\mathbb{R})$.

Exercise 2.1 Verify that the so-called **bell function** is an element of $C_0^\infty(\mathbb{R})$:

$$\phi(x) := \left[\begin{array}{ll} \exp[\frac{1}{x^2-1}], & |x| < 1 \\ 0, & |x| \geq 1 \end{array} \right. .$$

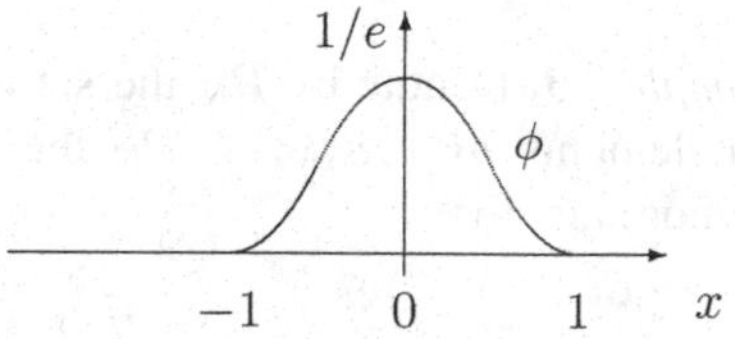

(*c*) Consider now the real infinitely differentiable functions $\phi(x)$, defined on the whole line and that satisfy:

$$M(\phi, k, p) := \max_{x \in \mathbb{R}} \left| x^p \frac{d^k \phi(x)}{dx^k} \right| < \infty \tag{2.1}$$

for any integers $k, p \geq 0$. We call them as **rapidly decreasing functions**. The set of these functions, associated with a convenient notion of convergence, is denoted by $\mathcal{S}(\mathbb{R})$ and known as **Schwartz space**. We will have chances to pay many visits to them later.

Exercise 2.2 Verify that $\psi(x) := \exp^{-x^2}$ and $\psi(p(x))$ belong to $\mathcal{S}(\mathbb{R})$, for any non-constant polynomial p.

2.2 Norm and Distance

It can be roughly expressed that functional analysis aim is to study vector spaces where we are able to measure **distances**. And, essentially, numerical analysis purpose is, in these environments, to **approximate** elements we are interested on by other ones whose access is easier. But a pattern must always be followed: to keep control upon the exchange prices. In a precise way, one ought to know the **approximation error** thereby introduced.

A **non-negative** function

$$n : V \to \mathbb{R}^+$$

defined on a real vector space V is said to define a **norm** if the following properties hold:

(i) $n(x + y) \leq n(x) + n(y) \forall x, y \in V$
(ii) $n(\alpha x) = |\alpha| n(x) \forall \alpha \text{real}, \forall x \in V$
(iii) $n(x) = 0 \Rightarrow x = 0$

Instead of $n(x)$, it is universally accepted to represent a norm by $\|x\|$. This concept, already familiar from two- and three-dimensional spaces, is interpreted as the **distance** from the vector x to the null vector 0, and this way the distance between two vectors x and y is given by $\|x - y\|$. Property i) is called **triangle inequality** due to its geometrical interpretation.

It turns out that on the same vector space, it is possible to deal with different norms. We shall verify that some issues, which are termed as **topological** properties of such spaces, may be kept or not, depending on the chosen norms.

Example 2.5 Consider in $\mathbb{R}^N$ the three norms that follow:

$$\|x\|_1 := \sum_{j=1}^{N} |x_j|, \tag{2.2}$$

$$\|x\|_2 := \sqrt{\sum_{j=1}^{N} |x_j|^2}, \tag{2.3}$$

$$\|x\|_\infty := \max_{1\le x\le N} |x_j|. \tag{2.4}$$

Exercise 2.3 Take $N = 2$ and draw the **unitary circles** that correspond to each one of these norms. In other words, get the graphs of

$$\{x \in \mathbb{R}^2;\ \|x\|_p \le 1\},\ \text{with } p = 1, 2, \infty.$$

Example 2.6 In ℓ^∞,

$$\|x\|_\infty := \sup_{j\in\mathbb{N}} |x_j| \tag{2.5}$$

defines a norm. In ℓ^2, we may consider, besides the norm (2.5),

$$\|x\|_2 := \sqrt{\sum_{j=1}^{\infty} |x_j|^2}. \tag{2.6}$$

And in ℓ_0^∞, it is possible to deal with, besides these two norms, a third one, namely:

$$\|x\|_1 := \sum_{j=1}^{\infty} |x_j|.$$

Example 2.7 In $\mathcal{P}_N$, for each fixed (and arbitrary) interval $[a, b]$, with $a < b$, we may consider the norm

$$\|p\| := \max_{a\le x\le b} |p(x)|.$$

Example 2.8 For $C^k[0, 1]$, we may deal with the following examples of norms:

$$\|f\|_1 := \int_0^1 |f(x)|dx,$$

$$\|f\|_2 := \sqrt{\int_0^1 |f(x)|^2 dx}, \tag{2.7}$$

$$\|f\|_\infty := \max_{a \le x \le b} |f(x)|,$$

or, more generally, being $0 \le r \le k$,

$$\|f\|_{r,1} := \sum_{J=0}^{r} \|d^J f/dx^J\|_1,$$

$$\|f\|_{r,2} := \sqrt{\sum_{J=0}^{r} (\|d^J f/dx^J\|_2)^2}, \tag{2.8}$$

$$\|f\|_{r,\infty} := \max_{0 \le J \le r} \|d^J f/dx^J\|_\infty.$$

Observe that, given a sequence $y = (y_n) \in \ell^2$, by defining the piecewise constant function f_y in $[0, \infty[$ by

$$f_y(x) := y_n, n \le x < n+1, n = 0, 1, \ldots,$$

the following identities hold

$$\|f_y\|_2^2 = \sum_{n=0}^{\infty} |y_n|^2 = \|y\|_2^2,$$

and in some sense this justifies employing the same notation for these norms in different spaces.

The main hardship when proving that the above definitions do lead to norms shows up while facing the triangle inequality, as regards to the norms $\|\cdot\|_2$. Nevertheless the proof gets smoothed out just by recalling the concepts in the section to be opened now.

2.3 Inner Product

By browsing through the $\mathbb{R}^3$ space, the concept of two-vector inner product becomes quite familiar. Recall its value amounts to the length of the projection of one of the vectors on the other, multiplied by the length of the latter, and affected by the sign of the cosine of the angle generated by both. Given an arbitrary real vector space, a real function

$$\begin{aligned} p : V \times V &\to \mathbb{R} \\ (x, y) &\to p(x, y) \end{aligned}$$

is considered to be an **inner product** whenever the constraints[1] below hold, for any choice of $x, y, z \in V$ and arbitrary $\lambda \in \mathbb{R}$:

$$\begin{array}{ll} \text{positive definiteness} & \left[\begin{array}{l} i)\ p(x,x) \geq 0 \\ \quad p(x,x) = 0 \Rightarrow x = 0 \end{array}\right. \\ \text{symmetry} & \quad ii)\ p(x,y) = p(y,x) \\ \text{bilinearity} & \left[\begin{array}{l} iii)\ p(x+y,z) = p(x,z) + p(y,z) \\ iv) \quad p(\lambda x, y) = \lambda p(x,y) \end{array}\right. \end{array}.$$

An inner product may get either of the representations $< x, y >$, $< x|y >$, (x, y), $(x|y)$, or even $x{\bullet}y$. In general, our option will be for $(x|y)$, but we will not give up of either of the remaining ones.

Provided an inner product is available, it may be taken as the gadget to introduce a norm. Just recall the geometrical interpretation of a general inner product described above. If fully justifies to define

$$\|x\| := \sqrt{(x|x)}. \tag{2.9}$$

In order to verify the triangle inequality, we must demonstrate the so-called **Cauchy–Buniakowski–Schwarz inequality**:

$$|(x|y)| \leq (x|x)^{1/2}(y|y)^{1/2}. \tag{2.10}$$

Once (2.10) is proven, we deduce that

$$\begin{aligned} (x+y|x+y) &= (x|x) + 2(x|y) + (y|y) \\ &\leq (x|x) + 2(x|x)^{1/2}(y|y)^{1/2} + (y|y) \\ &= \left[(x|x)^{1/2} + (y|y)^{1/2}\right]^2, \end{aligned}$$

and this assures that $\|\cdot\|$, as defined in (2.9), is indeed a norm. Let us prove (2.10). For t an arbitrary real number, and for any choice of x and y in V, it can be verified to be valid the inequality that follows:

$$(x+ty|x+ty) \geq 0.$$

It is then deduced that:

$$0 \leq (y|y)t^2 + 2(x|y)t + (x|x),$$

from which it may be seen that

[1] Note that, as long as symmetry holds, linearity with relation to one of the variables implies that the same holds for the other one, and thus we have bilinearity.

$$4(x|y)^2 - 4(y|y)(x|x) \leq 0,$$

or

$$|(x|y)| \leq (x|x)^{1/2}(y|y)^{1/2}.$$

Exercise 2.4 Verify that the norms introduced by (2.3), (2.6), (2.7), and (2.8) were defined from an inner product.

A vector space where an inner product holds is called **Euclidean**. It is important to keep in mind that not all norms derive from an inner product. Those ones that do show up as very convenient, with respect to different concepts. This fact is a consequence of being the geometry of Euclidean spaces strongly similar to that of $\mathbb{R}^3$. They do not "allow" the presence of strange elements like the balls described in Exercise 2.3.

Exercise 2.5 Verify: in any Euclidean space, for arbitrary v, w, we have

$$\|v + w\|^2 + \|v - w\|^2 = 2\|v\|^2 + 2\|w\|^2. \tag{2.11}$$

The geometrical interpretation of this identity justifies it being referred as the **parallelogram rule.**[2]

2.4 Convergence

A sequence $(x_n)_{n\in\mathbb{N}}$ of elements from V is said to converge if, for some $x \in V$, we have

$$\lim_{n\to\infty} \|x_n - x\| = 0. \tag{2.12}$$

We will call such element x as **limit of the sequence** $(x_n)_{n\in\mathbb{N}}$; the triangle inequality implies uniqueness for such a limit. We will also express this fact by saying that (x_n) approximates x, or that x is approximated by the sequence (x_n).

All operations with limits we are used to for real numbers stay valid on V. Suppose that $\lim_n x_n = x$, we then have:

[2] It may be proved that if (2.11) holds in a normed space V, for any vectors v, w, the norm in V is related to some inner product.

$$\left.\begin{array}{l} a.\ \lim_n \alpha_n = \alpha \Longrightarrow \lim_n \alpha_n x_n \quad = \alpha x \ (\forall \alpha \in \mathbb{R}) \\ b.\ \lim_n y_n = y \Longrightarrow \lim_n (x_n + y_n) = x + y \\ c.\ \lim_n y_n = y \Longrightarrow \lim_n (x_n | y_n) \quad = (x|y) \end{array}\right] . \tag{2.13}$$

We point out that this is just one of the convergence notions we will deal with. It is named **strong convergence** or **norm convergence**. Whether a sequence converges – or fail to – does not depend only on the sequence itself but also on the norm under consideration.

Exercise 2.6 Verify that, no matter the above norms introduced in $\mathbb{R}^N$ through expressions (2.2) to (2.4), we have

$$\|x_n - x\| \to 0 \Longleftrightarrow x_n^j - x^j \to 0,$$

where x_n^j and x^j, $j = 1, \ldots, N$, are the components of x_n and x, respectively.

Exercise 2.7 A sequence

$$(f_n)_{n \in \mathbb{N}} \text{ in } C^0[0, 1]$$

converges to

$$f \in C^0[0, 1]$$

with respect to the norm $\|\cdot\|_\infty$ if and only if

$$f_n \to f$$

uniformly. (The norm $\|\cdot\|_\infty$ is called **uniform convergence norm**.)

Exercise 2.8

(*a*) The sequence $(f_n)_{n \in \mathbb{N}}$ from $C^0[0, 1]$ defined by

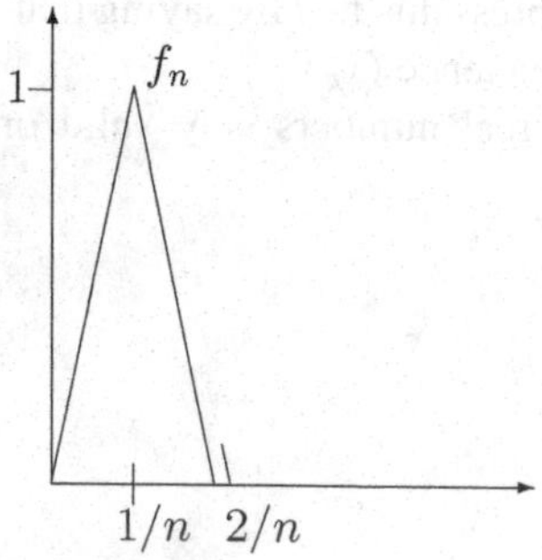

$$f_n(x) := \begin{bmatrix} nx & 0 \le x \le 1/n \\ 2 - nx & 1/n \le x \le 2/n \\ 0 & 2/n \le x \le 1 \end{bmatrix}$$

converges to $f \equiv 0$ with respect to the norm $\|\cdot\|_2$, that is, according to **quadratic average**, but not in the sense of the norm $\|\cdot\|_\infty$.

(*b*) The converse of *a*) never holds, since the limit $\|f_n - f\|_\infty \to 0$implies $\|f_n - f\|_2 \to 0$, provided all functions f_n live on a bounded domain.

Exercise 2.9 Both norms $\|\cdot\|_\infty$ and $\|\cdot\|_2$ may be considered on the space $C^0(\mathbb{R})$. Recall the bell function $\phi(x)$ from Exercise 2.1 to define the sequence

$$\psi_n(x) := \frac{\phi(x/n^2)}{n}$$

and then verify that $\|\psi_n\|_\infty \to 0$but$\|\psi_n\|_2$does not tend to0.

As previously remarked, our main purpose is to **approximate functions** or, in short, to build sequences that converge to an *a priori* presented function. But, as already stated, the notion of convergence is strongly associated to the chosen norm on our working function space. It is therefore natural to ask: How do these convergence notions compare themselves with that one which is more familiar to everybody, namely, **pointwise convergence**?

Recall that, for an arbitrary set X, being

$$\phi_n : X \to \mathbb{R}, n = 1, 2, \ldots,$$

a given sequence of functions, we say that ϕ_n shows **pointwise convergence** to ϕ on X whenever

$$\lim_n \phi_n(x) = \phi(x), \forall x \in X.$$

Beware that such definition still makes sense when the functions ϕ_n have as their range a normed space V.

As already remarked, convergence in $C^0[0, 1]$ relative to the norm $\|\cdot\|_\infty$ amounts to uniform convergence, and therefore it implies pointwise convergence as well as convergence with respect to quadratic average. On the other hand, quadratic average convergence does not imply pointwise convergence, neither the latter implies the previous one. The two following examples indicate these facts.

Example 2.9 Consider

$$h(x) := \left[\begin{array}{r} 2x \; x \in [0, 1/2] \\ 2(1-x) \; x \in [1/2, 1] \\ 0 \; x \notin [0, 1] \end{array} \right.$$

and

$$h_n(x) := h(2^{k_n} x - m_n),$$

where

$$\left.\begin{array}{l} k_n := \max_{\ell \in \mathbb{N}}\{2^\ell \leq n\} \\ m_n := n - 2^{k_n} \end{array}\right] .$$

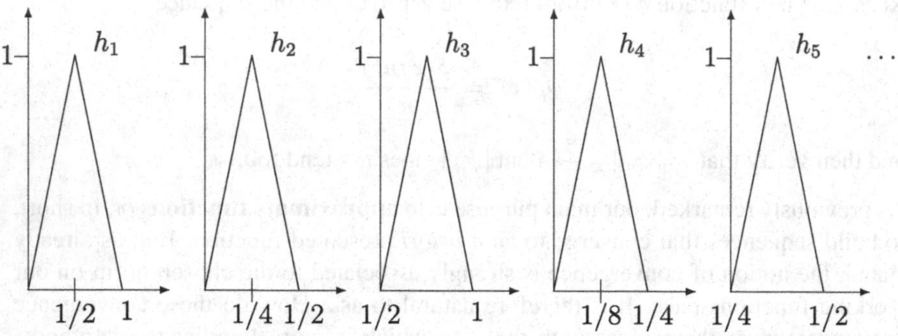

For an infinite amount of points all written as $2^{\imath+1}/2^N$, with $\imath = 0, 1, \ldots, N-2$, the functions h_n have their alternating values, either 0 or 1. Consequently, this sequence fails to show pointwise convergence. Nevertheless, just calculate the integrals $\int h_n^2$ to deduce the convergence of (h_n) to the null function, as regards to the quadratic mean.

Example 2.10 Take the sequence

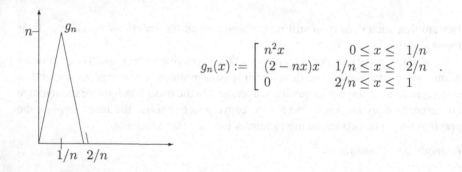

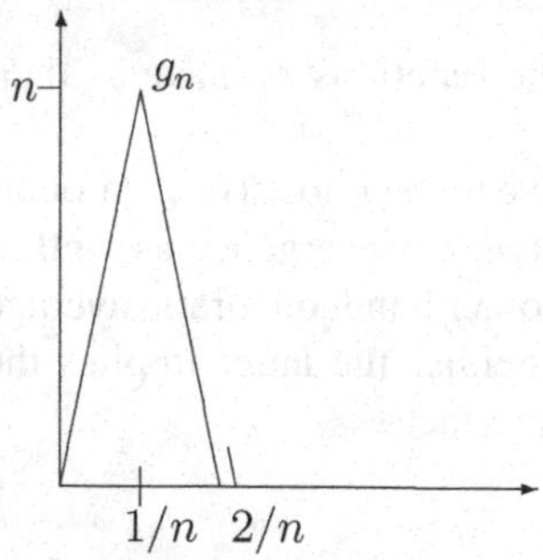

$$g_n(x) := \left[\begin{array}{ll} n^2 x & 0 \leq x \leq 1/n \\ (2 - nx)x & 1/n \leq x \leq 2/n \\ 0 & 2/n \leq x \leq 1 \end{array}\right. .$$

This sequence exhibits pointwise convergence to the null function in [0,1], but we verify that $\int g_n^2$ does not converge to zero.

Example 2.11 Important consequences are driven from the fact that, in $C^1[0, 1]$, convergence with respect to the norm $\|\cdot\|_{1,2}$ implies convergence relatively to the norm $\|\cdot\|_\infty$. Indeed, for $f \in C^1[0, 1]$, we have that

$$\int_0^1 |f(s)|ds = |f(x_0)|$$

for some $x_0 \in [0, 1]$. Now, for any x in $[0, 1]$, the identity

$$f(x) - f(x_0) = \int_{x_0}^{x} f'(s)ds$$

holds; thus it follows that

$$|f(x)| \leq \int_0^1 |f(s)|ds + \int_0^1 |f'(s)|ds.$$

By making use of Schwarz inequality (in fact, of CBS, expression (2.10)),

$$|\int_0^1 \phi(x)\psi(x)dx| \leq \|\phi\|_2\|\psi\|_2$$

with $\psi := 1$ and $\phi := f$ or f', it follows that

$$|f(t)| \leq \|f\|_2 + \|f'\|_2.$$

Therefore,

$$\begin{aligned} \|f\|_\infty^2 &\leq \|f\|_2^2 + \|f'\|_2^2 + 2\|f\|_2\|f'\|_2 \\ &\leq 2\{\|f\|_2^2 + \|f'\|_2^2\} = 2\|f\|_{1,2}^2 \end{aligned} \tag{2.14}$$

holds, since $2|ab| \leq a^2 + b^2$.

2.5 Continuous Functions

Given two normed spaces V, W, a function $\phi : V \to W$ is said to be **continuous** if, for any convergent sequence (v_n) in V,

$$\lim_{n\to\infty} \phi(v_n) = \phi(\lim_{n\to\infty} v_n).$$

Example 2.12

(*a*) No matter which norm we would be dealing with in $\mathbb{R}^n$, we can anticipate the continuity of the n projections

$$\left.\begin{array}{rll}\delta_j : & \mathbb{R}^n & \to \mathbb{R}, \\ & x = (x_1, \ldots, x_n) & \to \delta_j(x) := x_j, 1 \leq x \leq n\end{array}\right] 1 \leq x \leq n;$$

(*b*) For all $0 \leq t_0 \leq 1$, we have the continuity of the functionals

$$\begin{array}{rl}\delta_{t_0} : C^0[0,1] \to \mathbb{R} & \\ f & \to \delta_{t_0}(f) := f(t_0)\end{array}$$

as regards to $\|\cdot\|_\infty$, but not when $\|\cdot\|_2$ is considered;

(*c*) No matter which vector w is fixed in a given Euclidean space V, the functional defined through

$$\begin{array}{rl}\phi_w : V \to \mathbb{R}, & \\ v & \to \phi_w(v) := (v|w)\end{array}$$

is always continuous with respect to the norm associated to its inner product, as a consequence of inequality (2.10);

(*d*) For any $G \in C^1(\mathbb{R}^2)$, it can be shown the continuity of

$$\left.\begin{array}{rl}\phi : C^0[0,1] \to C^0[0,1] & \\ f & \to [\phi(f)](t) := \int_0^t G(s, f(s))ds\end{array}\right],$$

with respect to the norm $\|\cdot\|_\infty$.

It is straightforward to verify that:

(*i*) $\phi_1, \phi_2 : V \to W$ continuous $\Rightarrow \phi_1 + \phi_2$ continuous
(*ii*) $\alpha \in \mathbb{R}, \phi : V \to W$ continuous $\Rightarrow \alpha\phi$ continuous

There is no need to define the continuity concept for only functions defined on the whole space, as above: their domain may be as well an arbitrary subset of V. Indeed, the formulation just introduced requires only to put the hands on the notion of a convergent sequence and consequently to deal with distance between elements in the space under consideration.

2.6 The Open, Closed, Dense Sets

A subset F from a normed space V is **closed** if the limit of any convergent sequence (v_n) of elements from F belongs necessarily to F. With alternate words, the elements in F may only **approximate** elements from F itself.

Chosen $v_0 \in V$ and $r > 0$, the **closed ball** with **center** in v_0 and **radius** r is the set

$$B[v_0; r] := \{v \in V; \|v - v_0\| \leq r\}.$$

Exercise 2.10 As a consequence of the triangle inequality, $B[v_0; r]$ is always a closed set.

a) Verify that in $C^0[0, 1]$, the set $\{f; |f(t)| \leq 1, 0 \leq t \leq 1\}$ is closed relative to the norm $\|\cdot\|_2$.

b) Verify that $\{x \in \ell^2; |x^\imath| \leq 1/\imath\}$ is closed in ℓ^2 with respect to the norm $\|\cdot\|_2$. (This set is named the **Hilbert cube**.)

Given $X \subset V$, let us consider the set of all elements from V that tolerate to be approximated by vectors in X. This set is called the **closure of** X and is denoted by $\overline{X}$. Rephrasing it: $x \in \overline{X}$ if and only if x appears as limit of some sequence (x_n) with all $x_n \in X$.

Exercise 2.11 Verify that, for $v_0 \in X, r > 0$, $B[v_0; r]$ is the closure of

$$B(v_0; r) := \{v \in V; \|v - v_0\| < r\}.$$

The set just defined in this Exercise is called the **open ball** with **center** in v_0 and **radius** r. A set $X \subset V$ is said to be **open** if its complement is closed, or in an alternate expression form, if its elements may only be approximated by vectors that live in X. Still another way to characterize this notion is described in the

Exercise 2.12 A set $X \subset V$ is open if and only if, for any $x_0 \in X$, there exists an open ball with center in x_0 and entirely contained in X. It can be verified that $B(v_0; r)$ is always open, for any choice of $r > 0$ and any $v_0 \in V$.

A subset $X \subset V$ is defined as **dense** (in V) if $\overline{X} = V$, *i.e.*, if every element of V may be approximated by elements of X.

Exercise 2.13 Verify that ℓ_0^∞ is dense in ℓ^2 with the norm $\|\cdot\|_2$ as well as in c_0 (with the norm $\|\cdot\|_\infty$), but it fails to be dense in c or in ℓ^∞.

Exercise 2.14 Is $C_0^\infty(\mathbb{R})$ dense in $\mathcal{S}(\mathbb{R})$ relative to the norm $\|\cdot\|_\infty$? What about choosing the norm $\|\cdot\|_2$?

An important sample of a dense set in $C^0[0, 1]$ is the space of all polynomials for either norm $\|\cdot\|_\infty$ or $\|\cdot\|_2$. In the first case, this is exactly the statement of the **Weierstrass Approximation Theorem** ([61], pp.146). The second claim follows from that result.

Exercise 2.15 Verify that the set of all polynomials with rational coefficients is as well dense in $C^0[0, 1]$, with respect to the uniform convergence norm. And what to say about L^2?

Dense sets may show, in particular events, that they carry more power than *a priori* thought: it turns out that knowledge of some data on determinate dense subsets suffices to make sure that these data also belong to the whole space under exam. As an example, suppose that $g \in C^0[0, 1]$ is such that

$$\int_0^1 x^r g(x)dx = 0 \tag{2.15}$$

for $r = 0, 1, \ldots$; does this guarantee that

$$\int_0^1 f(x)g(x)dx = 0 \tag{2.16}$$

for an arbitrary $f \in C^0[0, 1]$?
Well, from (2.15) we deduce that (2.16) holds whenever f is a polynomial. Take into account that the functional

$$f \to \int_0^1 f(x)g(x)dx$$

is continuous with regard to the norm $\|\cdot\|_\infty$. Choose then a sequence of polynomials (p_n) which approximates f, also according to this norm, in order to conclude that (2.16) holds for any $f \in C^0[0, 1]$.

Exercise 2.16 Conclude that $g \equiv 0$. •

By the same token, as long as it is known that $g : \mathbb{R} \to \mathbb{R}$ is continuous and vanishes on the set of rationals, we can conclude that $g \equiv 0$.

Remark that the two exercises above unravel a quite general scenario: a continuous function $f : V \to W$ is characterized by its values in any dense set X in V. In other words,

$$\left.\begin{array}{l} f_1, f_2 : V \to W \text{ continuous} \\ f_1(x) = f_2(x), \forall x \in X \\ X \text{ dense in } V \end{array}\right] \Longrightarrow f_1 \equiv f_2.$$

But this result, as a gift, raises a question: suppose that f is continuously defined on X which is dense in V. May it be assured that f holds a **continuous extension** to all of V? Another formulation: does it exist

$$F : V \to W$$

continuous and such that

$$F(x) = f(x) \text{ for all } x \in X?$$

It is known by now that, as long as such extension F exists, it is unique. In this case, it is said that F is the **continuous extension** for f. It turns out that this search can meet a blocking, though. For example, the function

$$\left. \begin{array}{lll} f : \{x \in \mathbb{R}; x \neq 0\} & \rightarrow & \mathbb{R} \\ x & \rightarrow & f(x) := \sin(1/x) \end{array} \right]$$

points out that such an extension may be forbidden to exist.

The first idea that pops up as one looks for a continuous extension, within such a framework, is to define, for all

$$x \in V \backslash X := \{x \in V; x \notin X\},$$

the values of the sought extension by

$$F(x) := \lim_n f(x_n),$$

having the sequence$x_n \in X$ the property: $x_n \rightarrow x$.

The adopted itinerary requires at once to answer: Is it indeed possible to assure these limits existence, besides – a foremost point – its independence of the chosen sequence (x_n)? We will indicate below which conditions on f and about the space W must be imposed.

2.7 The Cauchy Sequences

The definition of a **convergent** sequence of vectors $(x_n)_{n \in \mathbb{N}}$ in a normed space V, cf. (2.12), is **extrinsic**, since it employs data that are **external** to the sequence under analysis, namely, a particular point that turns out to be its limit. Nevertheless, it turns out that either this limit is unknown, and thus its value is being sought, or even its very existence is unknown, or, last but not least, it needs to be **approximated**. This is the strong endorsement to introduce the so-called **Cauchy criterium**.
For any convergent sequence $(x_n)_{n \in \mathbb{N}}$, it follows from the triangle inequality that:

Given an arbitrary real $\epsilon > 0$, there exists a value $M = M(\epsilon)$ such that, if $n, m > M$, then

$$\|x_n - x_m\| < \epsilon.$$

Cauchy sequences are defined as those that satisfy such criterium.

It is rather clear: **every convergent sequence must be a Cauchy sequence**. It would be quite convenient to have its reciprocal to hold, since the Cauchy criterium formulation deals just with the elements of the sequence under exam, being thus **intrinsic**.
Let get this point illustrated by considering the function $\theta \in C_0^\infty(\mathbb{R})$ defined by in order to construct the sequence

$$\psi_n(x) := \theta(x/n) \exp^{-x^2}.$$

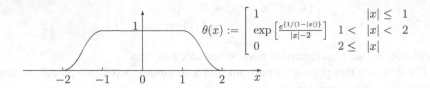

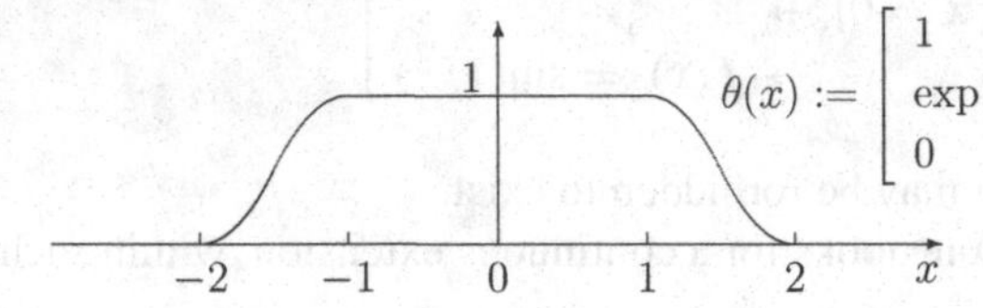

$$\theta(x) := \begin{bmatrix} 1 & & |x| \leq 1 \\ \exp\left[\frac{e^{\{1/(1-|x|)\}}}{|x|-2}\right] & 1 < & |x| < 2 \\ 0 & 2 \leq & |x| \end{bmatrix}$$

Exercise 2.17 Verify: the sequence $\{\psi_n\}$ is a Cauchy sequence in $C_0^\infty(\mathbb{R})$ under the norm $\|\cdot\|_\infty$, though it fails to be convergent, within this framework.

The same flavor comes from the example which follows: consider in ℓ_0^∞

$$x^n = (x_j^n)_{j\in\mathbb{N}} \text{ with } x_j^n := \begin{bmatrix} 1/j & j \leq n \\ 0 & j > n \end{bmatrix}.$$

Exercise 2.18 Prove that (x^n) is a Cauchy sequence in ℓ_0^∞, no matter if the norm $\|\cdot\|_\infty$ is taken or $\|\bullet\|_2$ is our choice; but it does not converge in ℓ_0^∞, under either norm.

A quite important fact is present in both examples. Observe that

$$C_0^\infty(\mathbb{R}) \subset \mathcal{S}(\mathbb{R}),\ \ell_0^\infty \subset \ell^2 \subset \ell^\infty.$$

The above considered sequences (ψ_n) and (x^n) belong, thus, to $\mathcal{S}(\mathbb{R})$ and ℓ^2, respectively. It turns out that, being

$$\psi(x) := e^{-x^2} \in \mathcal{S}(\mathbb{R}),$$

$$x = (x_j)_{j\in\mathbb{N}} := (1/j)_{j\in\mathbb{N}} \in \ell^2 \subset \ell^\infty,$$

we can claim that

$$\psi_n \to \psi \text{ in } \mathcal{S}(\mathbb{R}),$$

under the norm $\|\cdot\|_\infty$, and

$$x^n \to x \text{ in } \ell^2 \text{ or } \ell^\infty.$$

Therefore, the Cauchy sequences under consideration lack their convergence inside the spaces first taken, but they become convergent provided we deal with the "correct" ones. The choice done at first has pointed to spaces not "rich" enough, as they lack the right elements to guarantee the sought convergence.

A **Banach space** is defined as a normed space in which any Cauchy sequence converges; on the other hand, an Euclidean space with this same property is defined as a **Hilbert space**.

All spaces $\mathbb{R}^N$, independently of the chosen norm, are examples of Banach spaces, while $C^0[0, 1]$ is a Banach space if we consider the norm $\|\cdot\|_\infty$, but lacks this property under $\|\cdot\|_2$. This last claim can be verified by considering, for example, for $n > 2$, the sequence defined as follows.

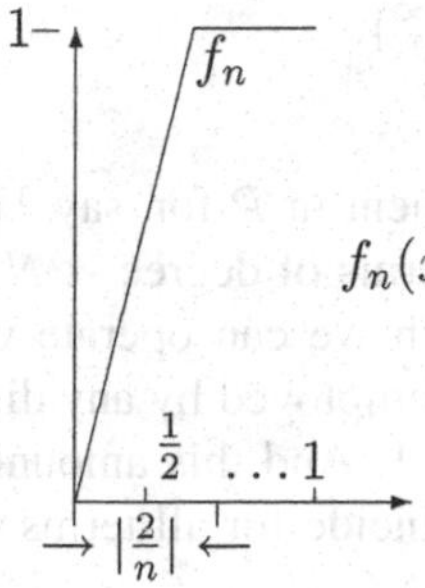

$$f_n(x) := \begin{bmatrix} 0 & 0 \le x \le \frac{1}{2} - \frac{1}{n} \\ n\left(\frac{x}{2} - \frac{1}{4}\right) + \frac{1}{2} & \frac{1}{2} - \frac{1}{n} \le x \le \frac{1}{2} + \frac{1}{n} \\ 1 & \frac{1}{2} + \frac{1}{n} \le x \le 1 \end{bmatrix}$$

Exercise 2.19 Prove that $C^1[0, 1]$ is complete under the norm $\|\cdot\|_{1,\infty}$, but it is not a Banach space under the norm $\|\cdot\|_\infty$.

Exercise 2.20 Verify that, for $1 \le p < \infty$, all the spaces

$$\ell_p := \{x = (x_j); x_j \in \mathbb{R} \text{ and } \sum_{j=1}^{\infty} |x_j|^p < \infty\}$$

are as well complete, as long as we choose the norms

$$\|x\|_p := \{\sum_j |x_j|^p\}^{1/p}.$$

Exercise 2.21 Demonstrate that, under the norm $\|\cdot\|_\infty$, we have completeness for the spaces ℓ_∞, c, and c_0.

Exercise 2.22 Provided $1 \le p < q \le \infty$, deduce that $\ell_p \subset \ell_q, \ell_p \ne \ell_q$.

2.8 Quotient Spaces

A concept that occupies an important slot in our theoretical development is that of **quotient spaces**, which arises as fruit of different frameworks, one of them described in

Example 2.13 Let $\mathcal{P}$ be the space of all real coefficient polynomials

$$\mathcal{P} := \{p(x) := \sum_{\imath=0}^{\infty} a_\imath x^\imath ; a = (a_\imath) \in \ell_0^\infty\}.$$

If we need to compute the values $p(\bar{x})$ of an arbitrary element in $\mathcal{P}$ for, say, $|\bar{x}| \leq 1/2$, we will necessarily have to restrict ourselves to the terms of degree $< N$, for some N which will depend on the number of digits which we can operate with. This is a commandment from the floating point arithmetic employed by any digital computer: $x^N = 0$ for N large enough, as long as $|x| < 1$. And this amounts to be unable do **distinguish** two distinct polynomials that coincide for all terms with degree $< N$, when we are working with values for $|x| < 1$.

Two polynomials p and q are considered to be **equivalent** if

$$T_N(p - q) = 0, \tag{2.17}$$

where T_N is the **truncation operator**

$$\begin{array}{lll} T_N : & \mathcal{P} & \to \mathcal{P}_N \\ & p(x) := \sum_{\imath=0}^{\infty} a_\imath x^\imath & \to (T_N p)(x) := \sum_{\imath=0}^{N-1} a_\imath x^\imath \end{array} .$$

The notations $p \sim q$ or $p R_N q$ are both assigned, and we have:

$$\begin{array}{llll} a. \text{ Reflexivity } p \sim p & & \forall p \in \mathcal{P} \\ b. \text{ Symmetry } p \sim q & \Longleftrightarrow q \sim p & \forall p, q \in \mathcal{P} \\ c. \text{ Transitivity } p \sim q, q \sim r & \Longrightarrow p \sim r & \forall p, q, r \in \mathcal{P} \end{array} \tag{2.18}$$

We express then that (2.17) defines an **equivalence relation**.

To each $p \in \mathcal{P}$, we associate its **equivalence class**

$$p^* := \{q \in \mathcal{P}; q \sim p\},$$

which amounts to the set of all polynomials that may have terms distinct from the corresponding ones of p only if these have degree $\geq N$. The collection of these equivalence classes, denoted by $\mathcal{P}/\sim$, is named **quotient set**.

The operations on a vector space are then introduced in $\mathcal{P}/\sim$:

$$\left.\begin{array}{ll} a) & p^* + q^* := (p+q)^* \\ b) & \lambda p^* := (\lambda p)^* \end{array}\right] . \tag{2.19}$$

At first sight, the right-hand terms in expressions (2.19) could depend on the chosen element that "represents" the equivalence class in consideration, and this way the operations thus introduced would fail to be **well defined** in $\mathcal{P}/\sim$. Making it more clear, consider, for example, (2.19.b). If $p_1 \sim p_2$, then $p_1^* = p_2^*$, and thus, $\lambda(p_1^*) = \lambda(p_2^*)$ could be defined by $(\lambda p_1)^*$ as well as by $(\lambda p_2)^*$, and we ought to reach the same element of $\mathcal{P}/\sim$. And, indeed,

$$p_1 \sim p_2 \Longrightarrow (\lambda p_1) \sim (\lambda p_2) \Longrightarrow (\lambda p_1)^* = (\lambda p_2)^*, \forall \lambda \in \mathbb{R}.$$

Exercise 2.23 Show that (2.18.a) effectively **defines** an addition in $\mathcal{P}/\sim$, which, with the operations we introduced, becomes a vector space.

In order to reach the space $\mathcal{P}/\sim$, called **quotient space**, the track followed is to identify elements associated to some property. It is a scenario that recalls the one with the **free vectors** on the three-dimensional space which are **identified** between themselves whenever on the same direction, orientation, and length.

Now, as long as

$$\mathcal{F}_N := \{p(x) := \sum_{\iota=N}^{\infty} a_\iota x^\iota ; (a_\iota) \in \ell_0^\infty \text{ with } a_\iota = 0 \text{ if } 0 \le \iota < N\},$$

then $0^* = \mathcal{F}_N$ and (2.17) may be rewritten

$$p \sim q \Longleftrightarrow p - q \in \mathcal{F}_N. \tag{2.17'}$$

The relation $\sim$ is defined by the operator T_N or else by its **kernel** $\mathcal{F}_N$. In a more general environment, given E, a vector space and $F \subset E$, a subspace,

$$x \sim_F y \Longleftrightarrow x - y \in F \tag{2.20}$$

defines always an **equivalence relation** – *i.e.*, fulfills the properties listed in (2.18) – and $E/\sim_F$ is a vector space when we introduce the operations in the same way as in (2.19).

Example 2.14 Let $E := \mathbb{R}^3$, $0 \neq y \in \mathbb{R}^3$ fixed and $F := Y$ be the line generated by y, *i.e.*, $Y := \{x \in \mathbb{R}^3 | x = \alpha y, \alpha \in \mathbb{R}\}$. Then, for every $x \in \mathbb{R}^3$, the equivalence class of x^* is the line which is parallel to Y and which meets x.

In this framework, it is possible to measure the distance between two classes x_1^*, x_2^*, which amounts to introduce the norm of $x^* :=$ distance of x to the line Y.

This example guides us to wonder whether is it possible to work with the above notions inside normed spaces. In a more precise formulation, being E a normed

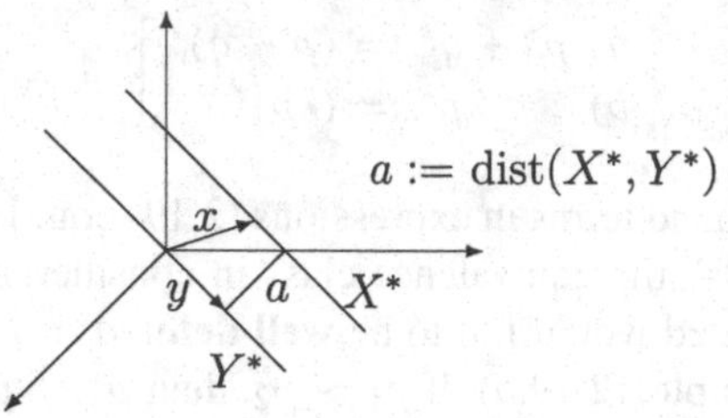

space and given one of its subspaces, say F, how to introduce a norm in $E/\sim_F$? Just observe that, in any normed space,

$$\|x\| = \text{distance from } x \text{ to the origin},$$

and that in $E/\sim_F$, $0^* = F$. In other words, F is the origin in the quotient space; therefore, it is quite natural to define

$$\|x^*\| := \text{distance from } x^* \text{ to } F, \tag{2.21}$$

provided we know how to introduce the notion of **distance between two sets**. Such a notion is defined then, for $A, B \subset E$ (= normed space), by means of:

$$\text{dist}\,(A, B) := \inf\{\|a - b\|;\, a \in A, b \in B\}.$$

The step that follows at once is to confirm that

$$\|x^*\| = \inf\{\|x - f\|;\, f \in F\}. \tag{2.21$'$}$$

It must be taken into account that, in the right-hand side of this expression, x may be replaced by any $y \in x^*$. This remark lets (2.21′) be taken as an **alternate definition**. Besides, it can be verified that the following identity holds:

$$\|x^*\| = \inf\{\|x\|;\, x \in x^*\}. \tag{2.21$''$}$$

One question stays on the waiting stop: Does (2.21) actually define a norm? Observe that no **topological** hypothesis was made about F. Unfortunately, if F lacks to be closed and $x \in \overline{F}$, but $x \notin F$, then $x^* \neq 0^* = F$, despite occurring that $\|x^*\| = 0$. And this happens since a sequence exists for which

$$f_n \in F,\, f_n \to x.$$

Now, this is exactly the missing hypothesis, as it can be seen from

Exercise 2.24

(*a*) Given a normed space N, if $F \subset N$ is a closed subset, (2.21′) defines a norm in $E/\sim_F$, where $\sim_F$ is introduced with (2.20).
(*b*) If N is a Banach space, so is $E/\sim_F$. And the converse?

2.9 Completion of a Normed Space

Throughout the examples in Sect. 2.7, convergence of a given Cauchy sequence would be obtained were the space considered at first replaced by a **larger one**. And it turns out that this is exactly the rationale backing what is called **completion of a normed space**. It amounts to a general strategy for constructing an environment where every Cauchy sequence from a given normed space becomes convergent – inside this brought-in framework.

Let V be an arbitrary normed space. We will borrow its structure so as to build a Banach space $\tilde{W}$ which, in a sense to be made precise, contains V as a dense subset. Denote by W the set of all Cauchy sequences (x_j) with x_j from V. This set W becomes a normed space with the definitions choice:

$$\left.\begin{array}{l} i)\ (x_j) + (y_j) := (x_j + y_j) \\ ii)\ \alpha(x_j) := (\alpha x_j) \\ iii)\ \|(x_j)\|_W := \lim \|x_j\|_V \end{array}\right] .$$

The triangle inequality guarantees that $i)$ effectively defines an element in W. This claim is quite clear for $ii)$. In order to prove that $iii)$ defines indeed a norm, we must at once verify existence of the limit in the right-hand side and, for that, to take hold of the inequality (2.22), which is important *per se*.
Let x, y be arbitrary vectors from a normed space. From the triangle inequality, it follows that

$$\|x\| = \|x - y + y\| \leq \|x - y\| + \|y\|,$$

which implies

$$\|x\| - \|y\| \leq \|x - y\|. \tag{A}$$

Now just exchange y and x in (A) so as to deduce that:

$$-\|x - y\| \leq \|x\| - \|y\|. \tag{B}$$

From expressions (A) and (B),

$$-\|x - y\| \leq \|x\| - \|y\| \leq \|x - y\|$$

follows, which amounts to

$$| \|x\| - \|y\| | \leq \|x - y\|. \tag{2.22}$$

(We should remark that this relationship was deduced by employing just the triangle inequality and that, for any vector x, it is true that $\| - x\| = \|x\|$.)
It turns out that the "function" introduced in W by means of iii) fails to be a norm. Indeed, the zero vector in W is

$$(x_j)_{j \in \mathbb{N}}, \text{ where } x_j = 0 \forall j \in \mathbb{N},$$

but $\lim_j \|x_j\|_V = 0$ implies only that (x_j) is a sequence that **approximates** the null vector in V.
We better retrieve the previous section where a norm fails to be defined by (2.21) in $E/\sim_F$, for not being **closed** in W the subspace F. Now we deal with a space that will show up as normed space only after an **identification** has been introduced between its elements; or, with a better saying, after being defined an **equivalence relation**. We find as more natural to treat this problem first in a more generic context and then step down to the particular case of W.

Let N be a vector space where a **semi-norm** $s(\cdot)$ has been defined. For this we mean a non-negative, homogeneous function for which the triangle inequality[3] holds. In alternate words,

$$s(x) = 0 \Longrightarrow x = 0$$

is the only property of a norm which s **may fail** to fulfill. In order to obtain, from N, a normed space, consider the subspace

$$F := \{x \in N; s(x) = 0\}.$$

In the quotient space $N/\sim_F$, we introduce

$$\|M\| := s(y), \forall y \in M, \text{ with } M \in N/\sim_F .$$

For all $y \in M$, $s(y)$ exhibits the same value, since

$$y_1, y_2 \in M \Longleftrightarrow y_1 - y_2 \in F \Longleftrightarrow s(y_1 - y_2) = 0$$

and, by the same reasoning already called for to reach (2.22),

$$|s(y_1) - s(y_2)| \leq s(y_1 - y_2).$$

This implies that s becomes indeed a norm in $N/\sim_F$.

---

[3] Hereby also mentioned as **sub-aditivity**.

Take now as N the space W of Cauchy sequences introduced above, denoted by $\widetilde{W}$ the quotient space or, else, $\widetilde{W}$ will be W after the identification of two Cauchy sequences $x = (x_j)$ and $y = (y_j)$ of elements from V whenever

$$\lim \|x_j - y_j\|_V = 0.$$

The notation $\|\cdot\|_\sim$ will be adopted for the norm of $\widetilde{W}$.
For any $v \in V$, the **constant sequence** $w := (w_j)$, with $w_j := v$ for all $j \in N$, belongs to W. The element $v \in V$ is associated in a quite natural fashion to the class $\tilde{w} \in \widetilde{W}$ determined by such an element $w \in W$, and we have that

$$\|\tilde{w}\|_\sim = \|v\|_V. \tag{2.23}$$

That is the meaning employed when it is said that $V \subset \widetilde{W}$. We shall employ the notation $\widetilde{V}$ for this subset of $\widetilde{W}$ which, from now on, we identify with the originally considered space V. Alternatively said, $\widetilde{V}$ is the set of the (classes of) sequences whose limit exists (inside V). Property (2.23) points out that the function

$$v \in V \rightarrow \tilde{v} \in \widetilde{V} \subset \widetilde{W}$$

preserves the norm or, with mathematical slang, it is an **isometry**.
Let us show that any $\tilde{w} \in \widetilde{W}$ may be approximated, with an arbitrary precision level, by elements $\tilde{v} \in \widetilde{V}$.
Let $(w_j) \in W$ be a "representative" of the given class $\tilde{w} \in \widetilde{W}$, and let $\epsilon > 0$ be an arbitrary real. Since (w_j) is a Cauchy sequence, there exists $N = N(\epsilon)$ such that, as long as $m, n \geq N$, we have

$$\|w_m - w_n\|_V < \epsilon.$$

It follows then that (v_j^ϵ) defined by $v_j^\epsilon := w_{N(\epsilon)}$, for any j, is associated to an element $\tilde{v}_\epsilon \in \widetilde{V}$ and that

$$\|\tilde{v}_\epsilon - \tilde{w}\|_\sim < \epsilon.$$

The proof that $\widetilde{W}$ is a Banach space follows now.
Let $(\tilde{w}_n)$ be a Cauchy sequence in $\widetilde{W}$. No matter the choice for n, it is possible to present an element $\tilde{v}_n \in \widetilde{V}$ for which $\|\tilde{v}_n - \tilde{w}_n\|_\sim < 1/n$. It is valid the claim that $(\tilde{v}_n)$ is a Cauchy sequence, since the chain of inequalities that follows is true:

$$\|\tilde{v}_n - \tilde{v}_m\|_\sim \leq \|\tilde{v}_n - \tilde{w}_n\|_\sim + \|\tilde{w}_n - \tilde{w}_m\|_\sim + \|\tilde{w}_m - \tilde{v}_m\|_\sim$$

$$\leq 1/n + 1/m + \|\tilde{w}_n - \tilde{w}_m\|_\sim.$$

Observe now that, given $\tilde{v} \in \widetilde{V}$, in a unique way it is associated to an element $v \in V$. This would then let us to consider the vectors $v_n \in V$ thus obtained from the corresponding $\tilde{v}_n$. This sequence (v_n) belongs to W (why is that so?), and we claim that it (would better be said, $\tilde{w}$, the class it determines inside $\widetilde{W}$) is the limit of $(\tilde{w}_n)$.
Indeed, we have

$$\|\tilde{w}_n - \tilde{w}\| \leq \|\tilde{w}_n - \tilde{v}_n\|_\sim + \|\tilde{v}_n - \tilde{w}\|_\sim \leq 1/n + \lim_j \|v_n - v_j\|_V \to 0,$$

which closes the proof.
This construction could have been carried out in no matter which **metric space**M, which means: a set where the notion of **distance** has been introduced. This means a function

$$\begin{aligned} d : M \times M &\to \mathbb{R}^+ \\ (x, y) &\to d(x, y) \geq 0 \end{aligned} \tag{2.24}$$

for which the following three assumptions may be claimed to hold:

$$\left.\begin{array}{l} d(x, y) = d(y, x) \\ d(x, y) = 0 \Longleftrightarrow x = y \\ d(x, z) \leq d(x, y) + d(x, z) \end{array}\right] .$$

A metric space where any Cauchy sequence necessarily converges is expressed as being **complete**.

Example 2.15 As remarked in Chap. 1, this general construction was motivated and guided by the case where

$$V = \mathbb{Q} := \text{ the set of rational numbers,}$$

for which we obtain

$$\widetilde{W} = \mathbb{R}.$$

Exercise 2.25 What are the output results $\widetilde{V}$ and $\widetilde{W}$ for an input space V already complete?

Example 2.16 The space $V := C^0[0, 1]$ with the norm $\| \cdot \|_2$ is not complete. Its completion turns out to be $\widetilde{W} :=$ set of real functions defined on $[0, 1]$, whose square are integrable on the Lebesgue sense. This fact may be taken as a road to define Lebesgue integral (see Sect. 2.14), or else may be proved, whenever this integral happens to be developed with alternate tools.

Example 2.17 With the choice of the space $C^1[0,1]$ and the norm $\|\cdot\|_{1,2}$ for V, we are lead to $\widetilde{W}$ as the Sobolev space $H^1(0,1)$, to be detailed on Chap. 4.

2.10 Principle of the Continuous Extension

Let's get back to the example (or counterexample) in Sect. 2.6.

Example 2.18 The function

$$\begin{aligned} f : \mathbb{R}\backslash\{0\} &\rightarrow \mathbb{R} \\ x &\rightarrow \sin(1/x) \end{aligned}$$

fails to own a continuous extension to the whole real line $\mathbb{R}$. This is a consequence of the existence of sequences of real numbers (x_n) that approach zero and such that $(f(x_n))$ is not convergent. The reason lies on the fact that, even being continuous the function f, as we take $\bar{x}$ closer and closer to the origin, its graph becomes wilder in a neighborhood of $\bar{x}$. The corresponding values of $|f'(x)|$ get larger and larger, pointing out that small changes for x lead to (proportionally) much larger variations in the values of $f(x)$. The same type of event is found with

$$\left[\begin{aligned} f : \mathbb{R}\backslash\{-1,0,1\} &\rightarrow \mathbb{R} \\ x &\rightarrow f(x) := 1/x(x^2-1) \end{aligned}\right].$$

Example 2.19 Name as f any of the functions $\cos x, \exp^{-|x|}$, or $\tan^{-1} x$ and observe that, given any real x_0, even with no information about $f(x_0)$, it is always possible to determine this value, just borrowing data from $f(x)$, for $x \neq x_0$. Let us rephrase it: were f defined only on $\mathbb{R}\backslash\{x_0\}$, it would exist exactly only one choice to define it in the whole line in such a way as it would become continuous – just take its very value $f(x_0)$.

What keeps these two examples apart may be explained by the following remark: whenever a Cauchy sequence (x_n) is chosen in the domain of the functions in Example 2.19, it is seen that $(f(x_n))$ is also a Cauchy sequence. Let this story be told with different words: these functions transform Cauchy sequences in Cauchy sequences. In contrast, functions in Example 2.18 do not satisfy this property.
A given function

$$f : D \subset M \rightarrow N$$

is said to be **uniformly continuous** whenever, given $\epsilon > 0$, there exists a counterpart $\delta = \delta(\epsilon) > 0$ (*i.e.*, a real positive δ, which depends **only** on the value of ϵ) and is such that we have

$$\|f(x) - f(y)\|_N < \epsilon \text{ as long } \|x - y\|_M < \delta, x, y \in D.$$

Here, M and N are arbitrary normed spaces, likewise as in

Exercise 2.26

(*a*) If $f : M \to N$ is uniformly continuous, then f preserves Cauchy sequences.
(*b*) Being $D \subset \mathbb{R}^n$ bounded, then $f : D \to N$ is uniformly continuous **if and only if** it preserves Cauchy sequences.

Exercise 2.27 If the partial derivatives $\partial f/\partial x_j$ of $f : D \subset \mathbb{R}^n \to \mathbb{R}$, for $j = 1, 2, \ldots, n$, are known to be bounded, then f is uniformly continuous.

Suppose now that D is a dense subset of a normed space M and let $f : D \to N$ be a uniformly continuous function. Had we assumed that f would be only continuous and had tried to **extend** f to any $x \in M \backslash D$ by means of the natural choice

$$f(x) := \lim_n f(x_n), \tag{2.25}$$

where $x_n \to x, x_n \in D$, such definition could fail to be a consistent one. Indeed, it would be quite possible to find distinct sequences (x_n) that would have the same element x as limit, but that would be mapped through f to sequences with different limits. As a matter of fact, it suffices to require from f to be uniformly continuous to assure that, as long as those limits **exist**, uniqueness holds. To prove that, let (x_n) and (y_n) be sequences in D which do converge to the same element $x \in M \backslash D$. The sequence (z_n) defined by

$$z_{2k} := x_k, z_{2k-1} := y_k, k = 1, 2, \ldots,$$

converges to x, which assures it as being a Cauchy sequence. From this token $(f(z_k))$ also happens to be a Cauchy sequence and as a consequence

$$\lim_k f(x_k) = \lim_k f(y_k).$$

How can one guarantee the existence of the limit in (2.25), no matter which $x \in M \backslash D$ is chosen? The hard point in this question does not lie within the space M nor the function f; rather, it is carried on by the space N: despite being $(f(x_n))$ Cauchy sequences, they may fail to be convergent in N. It is then clear that if N turns out to be a Banach space, we will have the right dressing on. It is then handy to express the above reasoning with the

Principle of the Continuous Extension – PCE. *Let D be a* **dense** *subset of a normed space M and let*

$$f : D \to N$$

be a **uniformly continuous** *function. If N is a* **complete** *space, there exists a* **unique** *continuous extension of f to the whole space M. Moreover, such extension preserves the uniform continuity.*

Example 2.20 Define in $C^1[0, 1]$ the function

$$\left.\begin{array}{ll}\delta_{\tilde{t}} : C^1[0,1] \to \mathbb{R} & \\ f & \to \delta_{\tilde{t}}(f) := f(\tilde{t})\end{array}\right],$$

where $\tilde{t} \in [0, 1]$ is fixed but arbitrary. When the norm $\|\cdot\|_{1,2}$ is chosen, it is seen that $\delta_{\tilde{t}}$ is continuous. In fact, suppose that $f_n \to f$. It can be verified, for $g \in C^1[0, 1]$, that

$$g(x)^2 \leq 2(\|g\|_2^2 + \|g'\|_2^2) = 2\|g\|_{1,2}^2, \forall x \in [0, 1],$$

from which it follows that

$$|f_n(\tilde{t}) - f(\tilde{t})| \leq \sqrt{2}\|f_n - f\|_{1,2},$$

and therefore the conclusion that $\delta_{\tilde{t}}$ is continuous is reached.

Exercise 2.28 Prove that $\delta_{\tilde{t}}$ is uniformly continuous.
We have previously mentioned that the completion of $C^1[0, 1]$ with the norm $\|\cdot\|_{1,2}$ gives birth to the Sobolev space $H^1[0, 1]$. From the remark just made in the previous exercise, we are able to conclude that the function $\delta_{\tilde{t}}$ owns a continuous extension to $H^1[0, 1]$. Moreover, since for the functions in $C^1[0, 1]$ one can deduce that[4]

$$\begin{array}{c}|f(t_1) - f(t_2)|^2 = \left|\int_{t_1}^{t_2} f'(s)ds\right|^2 \leq \\ \int_{t_1}^{t_2} |f'(s)|^2 ds \cdot |t_1 - t_2| \leq |t_1 - t_2| \cdot \|f\|_{1,2}^2,\end{array} \tag{2.26}$$

we can idealize the elements from $H^1[0, 1]$ as being **uniformly continuous** functions.

2.11 The Linear Operators

Given a function $f : X \to Y$, it is natural to ask which properties from the set X get preserved by f. Recall that we have deduced that uniformly continuous functions **preserve Cauchy sequences**. When dealing with completion, it was observed that an element $v \in V$ is associated to another element $\tilde{v} \in \tilde{V} \subset W$ for which

[4] In (2.26), the first inequality may be obtained as a consequence of (2.42), or even from CBS (2.10).

$$\|\tilde{v}\|_{\sim} = \|v\|_V$$

holds, and thus such association **preserves** norms. Naturally connected to vector spaces is the next subject of our study, namely, linear transformations or operators:

$$T : V \to W,\ V, W \text{vector spaces},$$

is defined as **linear** if

$$T(\alpha_1 v_1 + \alpha_2 v_2) = \alpha_1 T(v_1) + \alpha_2 T(v_2)$$

for any choice of $v_1, v_2 \in V, \alpha_1, \alpha_2 \in \mathbb{R}$.
Traditionally, for linear operators, parentheses are omitted, so that we write Tv replacing $T(v)$.
In finite dimensional spaces, linear operators are naturally linked to matrices. All above quoted examples illustrate linear operators, with the exception of

$$f \to f(x_0) + \int_{x_0}^{x} G(t, f(t))dt,$$

which becomes linear only if G is linear with respect to the second variable. It is worth remarking the linearity which holds for the identification

$$\begin{aligned} \imath : V &\to \tilde{V} \subset \widetilde{W} \\ v &\to \tilde{v} \end{aligned}$$

of a space with a dense set in its completion.

Example 2.21 Let us quote the equation

$$u_{tt} = u_{xx}, t > 0, 0 \le x \le 1, \tag{2.27}$$

which models the free transverse vibrations of a flexible string. Take the boundary conditions

$$u(0, t) = u(1, t) = 0 \tag{2.27'}$$

that describe the string extreme points as fixed. It is known that, given arbitrary ϕ e ψ, provided they are regular enough,[5] it is possible to uniquely determine $u = u(x, t)$ that fulfills (2.27)–(2.27′), besides the initial conditions[6]

---

[5] – By this we mean that these functions are as differentiable as the calculations thereby required ask for –.

[6] In short, existence and uniqueness hold for this problem.

$$\left.\begin{aligned} u(x,0) &= \phi(x) \\ u_t(x,0) &= \psi(x) \end{aligned}\right] . \tag{2.27''}$$

Therefore, it is possible to define, for each fixed t_0, the operator

$$\{\phi, \psi\} \to u(\cdot, t_0) \in C^0[0,1],$$

where u represents the system **state** at t_0, which means the string displacement for $t = t_0$, or else, the value of the solution to (2.27)–(2.27″) at the instant t_0. It is possible then to verify – due to the linearity of the differential equation, as well as of the initial and boundary conditions, besides the uniqueness of the problem solution – that the considered operator is linear.

At this point it is natural to question: Which is the relationship between linearity and continuity?

Assume as continuous the linear operator T at a particular vector v_0. No matter which other vector v is chosen, if a sequence $v_n \to v$, it may be deduced that $(v_n + v_0 - v) \to v_0$ and therefore, being T continuous at v_0,

$$T(v_n + v_0 - v) \to T v_0.$$

Linearity implies then

$$T(v_n + v_0 - v) = T v_n + T v_0 - T v,$$

so that

$$T v_n - T v \to 0 \text{ or } T v_n \to T v.$$

This way, to verify whether is T continuous, it suffices to verify its continuity at a single point v_0, no matter the choice done. Such a condition may be expressed by:

$$T \text{ (assumed linear) is continuous in the whole space } V.$$
$$\Longleftrightarrow T \text{ it is continuous at the origin } \{v = 0\}.$$

Example 2.22 Consider

$$\left.\begin{aligned} T : C^1[0,1] &\to C^0[0,1] \\ f &\to Tf := f' \end{aligned}\right] .$$

Take in both spaces the norm $\|\cdot\|_\infty$. Verify then that T lacks continuity. Indeed, by choosing $f_n(x) := (1/n)\sin nx$, $\|f_n\| \le 1/n \downarrow 0$, but $f_n'(x) = \cos nx$, and consequently $\|f_n'\|_\infty = 1$.

It must be made clear by now that the notion of continuity depends on the norms thereby considered. If $C^1[0, 1]$ is equipped with the norm $\|\cdot\|_{1,2}$, while $C^0[0, 1]$ gets the norm $\|\cdot\|_2$, the operator T in Example 2.19 would turn out to be continuous.
Another result on linear operators follows.

$$T \text{ is continuous} \iff T \text{ is bounded}$$

or, *T maps balls inside balls:*

$$\{Tx; \|x\| \leq r_1\} \subset B(0; r_2) \text{ with } r_2 = r_2(T, r_1).$$

To prove it, suppose T is continuous, and let (x_n) be a sequence of elements from the ball $B(0; r_1)$. We claim that (Tx_n) is bounded since, if not, it would exist a subsequence [7] x_{n_k} with $x_{n_k} \neq 0$ and $\lim_k \|Tx_{n_k}\| = +\infty$. By taking

$$y_k := x_{n_k}/\|Tx_{n_k}\|$$

we get $y_k \to 0$ but $\|Ty_k\| = 1$, which contradicts the continuity of T.
In analogous fashion, suppose T to be bounded, but not continuous. It would exist, then, a sequence (x_n), $x_n \to 0$ and so that Tx_n does not tend to zero. This implies the existence of a subsequence (x_{n_k}) such that $\|Tx_{n_k}\| \geq \rho$, for some $\rho > 0$. The sequence $y_k := x_{n_k}/\|x_{n_k}\|$ is bounded, in spite of $\|Ty_k\| \to \infty$. These two claims can not stand together with the assumed boundedness for T.
For any linear and continuous operator T, we denote

$$|||T||| := \sup_{x \neq 0} \|Tx\|/\|x\| = \sup_{\|x\|=1} \|Tx\| = \sup_{\|x\| \leq 1} \|Tx\|, \tag{2.28}$$

from which it follows, for any $x \in V$,

$$\|Tx\| \leq |||T||| \cdot \|x\|,$$

where the notation leaves as implicit, as long as no doubt shows up, that

$$\|x\| := \|x\|_V \quad \text{and} \quad \|Tx\| := \|Tx\|_W.$$

Example 2.23 Let the operator

$$T : \mathbb{R}^N \to V$$

[7] We say that $\{x_{n_k}\}$ is a **subsequence** from $\{x_n\}$ if $\{n_k\}$ is a **strictly increasing** sequence.

be linear, being V an arbitrary normed space and $\{e_1, \ldots, e_N\}$ the canonical basis for $\mathbb{R}^N$. If

$$v_\iota := Te_\iota \text{ and } \|x\|_1 \leq 1,$$

$$\|Tx\|_V = \|T \sum \alpha_\iota e_\iota\|_V = \|\sum \alpha_\iota v_\iota\|_V$$

$$\leq \max \|v_\iota\|_V \sum |\alpha_\iota| = \max \|v_\iota\|_V \|x\|_1 = \max \|v_\iota\|_V.$$

This chain of deduced expressions implies that

$$|||T||| \leq \max \|v_\iota\|_V. \tag{2.29}$$

Exercise 2.29 Show that equality holds in (2.29).

Example 3A Just like in Example 2.23, it may be proven that

$$\|x\|_\infty \leq 1 \Longrightarrow \|Tx\| \leq N \max \|v_\iota\|_V, \tag{2.29'}$$

or else that

$$\|Tx\| \leq \sum \|v_\iota\|_V; \tag{2.29''}$$

provided $\|x\|_2 \leq 1$, we obtain

$$\|Tx\| \leq \sum |\alpha_\iota| \|v_\iota\|_V \leq (\sum \alpha_\iota^2)^{1/2} (\sum \|v_\iota\|_V^2)^{1/2} \leq (\sum \|v_\iota\|_V^2)^{1/2}. \tag{2.29'''}$$

Exercise 2.30 Consider the expressions (2.29′), (2.29″), and (2.29‴) and prove the existence of some vectors for which equality holds for every one of them.

The just exhibited example is in fact a particular show off of a general situation, namely:

Given any linear operator defined on a finite dimensional vector space, such operator is always continuous, ***independently*** *of the chosen norms, either on its domain or range* (see Sect. 2.13).

Definition Given a continuous linear operator T, the value $|||T|||$, introduced in (2.28), is called the **norm** of T **as operator** or, by now, its "norm."

Exercise 2.31 Verify that

$$\begin{aligned}|||T||| &= \inf\{\alpha \in \mathbb{R};\ \|Tx\| \le \alpha\|x\|, \forall x\}\\ &= \text{diameter } \{Tx;\ \|x\| \le 1\}.\end{aligned}$$

The "norm" of T is consequently the expansion – or contraction – factor for the unity ball when subjected to the action of T. In short, what we reached above was the following:

Theorem 2.1 *Let $T : M \to N$ be a linear operator which maps some normed space to another. All the claims that follow are then equivalent:*

(a) T is continuous at the origin.
(b) T is continuous at some point $x \in M$.
(c) T is continuous at any point $x \in M$.
(d) T is uniformly continuous.
*(e) T is a **Lipschitz function**, i.e., there exists a constant K for which*

$$\|Tx - Ty\|_N \le K\|x - y\|_M \forall x, y \in M;$$

(e′) There exists a real K, which depends on T, for which the inequality

$$||Tx|| \le K||x|| \tag{2.30}$$

holds, no matter which vector $x \in M$ is chosen.
(f) T maps the unity ball from M inside some ball contained in N.
(g) The image of any ball from M is contained in some ball in N.
(h) T maps bounded sets from M into bounded sets in N.

*Due to properties h) and e′), linear continuous operators have earned the identification of **bounded operators**.*

The following reading of properties a) to c) is more attractive: no matter which linear operator is considered, either it is continuous at some particular point from its domain – and therefore it turns out to be everywhere continuous – or else it is everywhere discontinuous. This reasoning makes it natural to recall the nowhere differentiable functions discussed on the first chapter. We can neither refrain from quoting the differentiability properties on Chap. 7.
Now let X and Y be two fixed normed spaces. Denote by $\mathcal{L}(X, Y)$ the set of all bounded linear operators from X to Y. As we take in $\mathcal{L}(X, Y)$ the "norm" defined by (2.28), it is verified that this space turns out indeed as a normed space. Thus, we can from now on write just norm, no quotes needed.
When considering $X = Y$, it is usual to employ $\mathcal{L}(X)$ instead of $\mathcal{L}(X, X)$.
Clearly for the identity operator$\mathbb{I} : X \to X$, we obtain

$$\mathbb{I} \in \mathcal{L}(X) \text{ and } |||\mathbb{I}||| = 1.$$

Furthermore, whenever

$$T \in \mathcal{L}(X, Y),\, S \in \mathcal{L}(Y, Z),$$

it may be deduced that

$$S \cdot T \in \mathcal{L}(X, Z) \text{ and } |||S \cdot T||| \le |||S||| \cdot |||T|||.$$

Exercise 2.32 Whenever Y is a Banach space, so is $\mathcal{L}(X, Y)$.

Example 2.2 Consider, for a fixed $\tilde{t} > 0$, the operator associated to the linear wave propagation equation

$$\left.\begin{array}{l} W_{\tilde{t}} : C_0^2[0,1] \times C^1[0,1] \to C^2([0,1] \times [0,\tilde{t}]) \\ \qquad\qquad \{\phi, \psi\} \longrightarrow u(x,t) := \text{solution of } (2.27)\text{–}(2.27'') \end{array}\right] .$$

In $\mathcal{V} := C_0^2[0,1] \times C^1[0,1]$, we consider the norm

$$\|\{\phi, \psi\}\|_{\mathcal{V}} := (\|\phi'\|_2^2 + \|\psi\|_2^2)^{1/2}$$

and in $Im(W_{\tilde{t}})$, the image of $W_{\tilde{t}}$ in $C^2([0,1] \times [0,\tilde{t}])$, the norm

$$\|u(x,t)\|_W := \max_{0 \le t \le \tilde{t}} (\|u_x(\cdot,t)\|_2^2 + \|u_t(\cdot,t)\|_2^2)^{1/2}.$$

(*Caveat*: When the **whole** space $C^2([0,1] \times [0,\tilde{t}])$ is considered, this expression no longer defines a norm.)

Given an arbitrary instant t_0, the so-called **system energy** at t_0 is introduced by means of

$$E(t_0) := \int_0^1 [u_x(x,t_0)^2 + u_t(x,t_0)^2] dx.$$

It can be checked that

$$\frac{d}{dt} E(t) = 2 \int_0^1 [u_{xt} u_x + u_{tt} u_t] dx = 2 \int_0^1 u_t [u_{tt} - u_{xx}] dx + 2 u_t u_x \,|_0^1 = 0,$$

where the last identity is a consequence of being null the integrand, due to (2.27), and as well null the boundary term, by (2.27′).
This reasoning has brought us to the **energy conservation**, which means that the system energy stays constant as time flows. From this property it follows, no matter which solution $u \in Im(W_{\tilde{t}})$ is under consideration, that

$$\|u(\bullet,\bullet)\|_W = \max_{0\le t\le \tilde{t}} \{E(t)\}^{1/2} = E(0)^{1/2} = \|\{\phi,\psi\}\|_{\mathcal{V}}$$

and this brings as a consequence:

$$|||W_{\tilde{t}}||| = 1.$$

Exercise 2.33 Calculate (or estimate) the norm of the linear operators described in Example 2.12 and Exercises 2.27 and 2.28.

2.12 Invertible Operators

Given the transformation $T : X \to Y$ between two normed spaces, the set $\{x|Tx = 0\}$ of all zeros of T is called the ***kernel*** of T, denoted by $ker(T)$. Whenever T is continuous, $ker(T)$ is closed. Moreover:

If T is linear, $ker(T)$ is a vector subspace.

This terminology lets a previous result to be rephrased as:

If T is continuous and its kernel contains a subset which is dense (with respect to the domain of T), then

$$T \equiv 0.$$

It is known that $0 \in ker(T)$ for linear T. Whenever $\{0\} = ker(T)$, then

$$Tx_1 = Tx_2 \Longrightarrow T(x_1 - x_2) = 0 \Longrightarrow x_1 - x_2 = 0 \Longrightarrow x_1 = x_2.$$

Thus the above assumption implies that T must be one to one, and thus it owns an inverse T^{-1}. Such an inverse ought to be linear. Would T^{-1} have to be continuous, under the assumption of being T continuous and linear?

Example 2.3 Let $\tilde{C}^1[0,1]$ be the set of all functions that belong to $C^1[0,1]$ and vanish on the origin, under the norm $\|\cdot\|_\infty$.

The operator

$$\begin{aligned} T : \tilde{C}[0,1] &\to \tilde{C}^1[0,1] \\ f \quad &\to (Tf)(x) := \textstyle\int_0^x f(s)ds \end{aligned}$$

is linear and continuous when chosen the norm $\|\bullet\|_\infty$ on $\tilde{C}[0,1]$, which denotes the space of functions from $C^0[0,1]$ that vanish at the origin. This is readily a consequence of the inequalities

$$|\int_0^x f(s)ds| \leq \|f\|_\infty \int_0^x ds \leq \|f\|_\infty.$$

Besides, the operator T is one to one since $f = 0$ if $Tf = 0$. With the definition

$$\left.\begin{array}{l} S : \tilde{C}[0,1] \to \tilde{C}[0,1] \\ \quad g \quad \to (Sg)(x) := g'(x) \end{array}\right],$$

S turns out to be the inverse of T, but it lacks continuity. To prove this fact, just take $g(x) := x^n$,

$$(Tg_n)(x) = \int_0^x t^n dt = \frac{x^{n+1}}{n+1},$$

from which it follows that

$$\|Tg_n\|_\infty = \max_{0\leq t\leq 1} \frac{t^{n+1}}{n+1} = \frac{1}{n+1} \downarrow 0.$$

This allows one to conclude that

$$\|g_n\|_\infty = \max_{0\leq t\leq 1} |t^n| = 1$$

and since

$$g_n = Sh_n \text{with} h_n(t) := t^{n+1}/(n+1),$$

we reach:

$$\|h_n\|_\infty \to 0 \text{ but } \|Sh_n\|_\infty = 1.$$

Therefore S is discontinuous. That lights the inspiration for the claims in

Exercise 2.34 Suppose given, for two arbitrary normed spaces, a linear and continuous function $T : X \to Y$. The inverse of T , which exists, defined throughout the image of T – hereby denoted by $Im(T) \subset Y$ – is continuous if and only if there exists a real $\delta > 0$ for which

$$\|x\| = 1 \text{ implies } \|Tx\| > \delta. \tag{2.31}$$

It is worth to keep an eye on the meaning of both expressions (2.30) and (2.31). In short, they bring estimates to the way the unit ball from M gets mapped to ∞ or away from the origin in N.

From a first look, being T linear, we should have

$$||Tx|| = \mathbf{O}(||x||)\text{as}||x|| \to \infty,$$

but as told by (2.30), this behavior is only allowed to bounded operators. On the other hand, Exercise 2.34 tells that only those invertible operators that map the unity ball boundary outside a fixed disk around the origin get the privilege to exhibit a continuous inverse.
An operator which is very often needed in applications for different subjects is the main actor in

Example 2.4 Let us denote by $\tilde{L}^1(\mathbb{R})$ the set of all functions $f : \mathbb{R} \to \mathbb{R}$ which are Riemann integrable in each closed and bounded interval and for which it is possible to guarantee the existence of the limit

$$\int_{-\infty}^{\infty} |f(x)|dx := \lim_{M,N\to\infty} \int_{-M}^{N} |f(x)|dx.$$

Take then the canonical vector operations as well as the norm

$$\|f\|_1 := \int_{-\infty}^{\infty} |f(x)|dx.$$

Now consider $C_a(\mathbb{R})$ as the set of continuous functions[8] $f : \mathbb{R} \to \mathbb{C}$ that vanish at infinity, *i.e.*,

$$\lim_{|x|\to\infty} f(x) = 0, \tag{2.32}$$

equipped with the norm

$$\|f\|_\infty := \sup_{-\infty<x<\infty} |f(x)|.$$

It can be deduced that the **Fourier transform**, defined through

$$\begin{array}{ll} \mathcal{F} : \tilde{L}^1(\mathbb{R}) \to C_a(\mathbb{R}) \\ \quad f \quad \to (\mathcal{F}f)(t) := \int_{-\infty}^{\infty} f(x)e^{-itx}dx \end{array}, \tag{2.33}$$

is a linear and continuous operator.[9]

---

[8] In the current framework, we must take hold of the **complex** vector spaces – CVS – for whose definition we just follow that one for a RVS, being the scalars replaced by complex numbers. While dealing with an inner product, comutativity is changed by

$$< u|v >= \overline{< v|u >}.$$

[9] The expression (2.32) describes a property of the Fourier transform known as the Riemann-Lebesgue lemma. For a proof, see [40], pp.303.

Exercise 2.35 Estimate the norm of $\mathcal{F}$ with as much precision as you can and verify that $\mathcal{F}$ does not have a continuous inverse. You may base your reasoning on facts revealed by the sequence

$$f_n(x) := \begin{bmatrix} 1 & |x| \le n \\ 0 & |x| > n \end{bmatrix}.$$

2.13 Equivalent Norms

We have already faced some doubts and conclusions that strongly depend on the norms employed in the spaces under consideration. Therefore one needs, whenever dealing with two different norms, to get hold of a handy criterion that would tell which information about one of them brings into play some information about the other one.

Example 2.5 Take the vector space $C^0[0, 1]$ and denote it by V_2 when the mean square norm is considered and by V_∞ if our choice goes to the norm $\|\cdot\|_\infty$. The identity operator

$$\begin{aligned} \mathbb{I} : V_2 &\to V_\infty \\ x &\to \mathbb{I}(x) = x \end{aligned}$$

is clearly linear. Pose the question: Is $\mathbb{I}$ continuous? To answer, it suffices to recall that the continuity of $\mathbb{I}$ would be equivalent to be a truth the following conclusion, which is false:

$$\text{convergence in } \|\cdot\|_2 \Longrightarrow \text{convergence in } \|\cdot\|_\infty.$$

Example 2.6 In $\tilde{C}^1[0, 1]$, $f \to \|f'\|_2$ defines a norm, which will be denoted by $\|\cdot\|_d$. It can be seen that the following inequality holds:

$$\|f\|_d \le \|f\|_{1,2}.$$

Since, for $x \in [0, 1]$,

$$f(x) = \int_0^x f'(s)ds, \; f \in \tilde{C}^1[0, 1],$$

we have

$$|f(x)|^2 \le \int_0^x |f'(s)|^2 ds \int_0^x 1 ds \le x \int_0^1 |f'(s)| ds$$

and thus

$$\int_0^1 |f'(x)|^2 dx \le \int_0^1 x dx \int_0^1 |f'(s)|^2 ds = \|f'\|_2^2/2.$$

This estimate, known as Wirtinger's or Friedrichs' inequality, is a special case of Poincaré inequality. One of its consequences is that, for $f \in \tilde{C}^1[0, 1]$:

$$\|f\|_{1,2}^2 = \|f\|_2^2 + \|f'\|_2^2 \le \frac{1}{2}\|f'\|_2^2 + \|f'\|_2^2 = \frac{3}{2}\|f'\|_2^2.$$

The pair of inequalities

$$\|f'\|_2 \le \|f\|_{1,2} \le \sqrt{3/2}\|f'\|_2$$

implies that the identity operator is continuous and has a continuous inverse when read as a mapping between the spaces

$$\tilde{C}^1([0, 1], \|\cdot\|_{1,2}) \text{ and } \tilde{C}^1([0, 1], \|\cdot\|_d).$$

We express that **two norms are equivalent** if whenever any sequence which is convergent relatively to one of these two norms is necessarily convergent with regard to the other one. Told in maybe a more precise fashion, $\|\cdot\|_1$ and $\|\cdot\|_2$ are equivalent norms on a space X, if two constants $\alpha, \beta > 0$ can be found such that

$$\alpha\|x\|_1 \le \|x\|_2 \le \beta\|x\|_1, \forall x \in X.$$

In a space where two equivalent norms were introduced, the convergent sequences and, as an aftereffect, all properties dependent to them turn out to be the same.

Let V be a real vector space. It is said that V is a finite dimensional space, with dimension N, $N \in \mathbb{N}$, and denoted $\dim(V) = N$, if there exists a linear transformation

$$T : \mathbb{R}^N \to V \tag{2.34}$$

which is both one to one and onto.

We get now to the proof of the already mentioned result:

Theorem *All norms introduced in a finite dimensional arbitrary normed space turn out to be equivalent.*

As told by Example 2.23, Sect. 2.11, the operator T from (2.34) is continuous, no matter which norm is the option among $\|\cdot\|_1, \|\cdot\|_2$, or $\|\cdot\|_\infty$. Our claim is that T^{-1} is continuous as well. This property rests proven as long as

$$\|Tx\|_V \geq \rho\|x\|_p \forall x \neq 0x \in \mathbb{R}^N, p = 1, 2, \infty,$$

for some $\rho > 0$, which amounts to

$$||Tx||_V \geq \rho \forall x \in \mathbb{R}^N, ||x||_p = 1,$$

or else

$$\inf_{||x||_p=1} ||Tx||_V > 0.$$

Taking into account that, if

$$\inf_{||x||_p=1} ||Tx||_V = 0$$

holds, there would exist a sequence $\{x_n \in \mathbb{R}^N, ||x_n||_p = 1\}$ for which

$$||Tx_n||_V \to 0.$$

Now, being the sequence $\{x_n\}$ is bounded, and as long as we have $x_n = (x_n^1, x_n^2, \ldots, x_n^N)$, this implies that each sequence $\{x_n^j\}_n, j = 1, \ldots, N$, is bounded as well. Such a fact implies to be possible then to extract off $\{x_n^1\}_n$ a convergent subsequence. When staying on this track, successively, component to component, we will reach a subsequence $\{x_{Jk}\}_k$ from $\{x_n\}$ in $\mathbb{R}^N$. We get convinced that convergence holds due to a previously quoted fact: a necessary and sufficient condition for any sequence in $\mathbb{R}^N$ to converge according to any one of the chosen norms $||\cdot||_p, p = 1, 2, \infty$, is that any of the sequences of reals formed by its components be convergent. In such a framework, we obtain that $x_{Jk} \overset{k}{\to} x_0$. Since $||x_n||_p = 1$, we have then

$$\lim_k ||x_{Jk}||_p = 1,$$

and this implies $\|x_0\|_p = 1$. From the continuity of T, it is deduced that

$$\lim_k Tx_{Jk} = T(\lim_k x_{Jk}) = Tx_0,$$

so that we conclude being $Tx_0 = 0$. But this contradicts to be T one to one.

From the continuity of T^{-1}, it follows equivalence of two **arbitrary** norms on any finite dimension space V. We suggest as an exercise to work out the remaining points in the proof.

Observations

(*a*) The main ingredient within the proof above is the Bolzano-Weierstrass Theorem which states:
Any bounded sequence in the real line (or in $\mathbb{R}^N$*) necessarily contains a convergent subsequence.*
The above described result belongs strictly to the finite dimension world, a point made explicit by the proof. Besides, a kind of reciprocal assertion holds, namely.

(*b*) *Assume a particular normed space V exhibits the property that follows: no matter which bounded sequence of its elements you are presented to, it is possible to extract from it a convergent subsequence. Such a space ought to be a finite dimensional one.*

Exercise 2.36 Verify that the set $\{\cos nx\}_{n\geq 0}$ is bounded in $C^0[0, 2\pi]$, with respect to the norm $\|\cdot\|_2$, but it owns no convergent subsequence. (Hint: Cauchy criterion.)

Example 2.7 In the space $\mathcal{P}_N$ of all polynomials with degree $< N$, consider both norms:

$$\left.\begin{array}{ll} ||p||_2 & := (\int_0^1 p^2(x)dx)^{1/2} \\ ||p||_{1,2} := & \left(\int_0^1 p^2(x)dx + \int_0^1 \left(\frac{dp}{dx}(x)\right)^2 dx\right)^{1/2} \end{array}\right] .$$

As it is known that the space $\mathcal{P}_N$ has dimension N,, the mentioned norms have to be equivalent. In other words, for each fixed N, there exist positive constants s_n and S_N such that, for each $p \in \mathcal{P}_N$,

$$s_N||p||_2 \leq ||p||_{1,2} \leq S_N||p||_2.$$

Existence of the constant s_n is quite clear ($s_n = 1$), but we can not state the same for S_N, if we are not aware of the above proven result. Within numerical analysis framework, the constant S_N is called **stability function** associated – in this case – to the pair $\{(\mathcal{P}_N, ||\cdot||_{1,2}), (\mathcal{P}_N, ||\cdot||_2)\}$.

Exercise 2.37 Verify that the conclusion below holds for the stability function:

$$S_N \uparrow \infty \text{if } N \uparrow \infty.$$

With the support of a result to be stated some pages forward,[10] we present

[10] It is indeed ranked as one of the three functional analysis core results, referred to as the **Open Mapping Theorem**.

Theorem 2.2 *Suppose that, on a vector space* V, *two norms are defined, namely,* $||\cdot||_1$ *and* $||\cdot||_2$. *Furthermore, with either one,* V *becomes a* ***complete*** *normed space. It can then be assured that the existence of* $K_1 > 0$ *for which*

$$||x||_1 \leq K_1||x||_2, \forall x \in V$$

is equivalent to the existence of $K_2 > 0$ *for which*

$$||x||_2 \leq K_2||x||_1, \forall x \in V.$$

In more intuitive terms, this theorem teaches that, under the assumptions it requires, if one of the norms dominates *the other, they show up as equivalent.*

2.14 Lebesgue Integral

2.14.1 Introduction

Lebesgue integral is a generalization of the Riemann integral. Associated with the former, there are some powerful results that do not hold for the latter, particularly some of which are related to convergence properties. The present section aims to discuss some basic facts within Lebesgue integral theory. They have shown themselves quite relevant (or even indispensable) for the applications we carry in mind. Unless pointed out, the functions dealt with throughout this section are all real, with some portion of the real line assigned as their domain, according to mentions conveniently made.

In a bounded[11] interval $[a, b]$, the class $\mathcal{R}[a, b]$ of bounded functions which are integrable in the Riemann sense – called as Riemann integrable – is **strictly** contained in the class of the functions which are integrable in Lebesgue sense, or Lebesgue integrable – $\mathcal{L}[a, b]$. (To get convinced of the term "strictly" written above, just look at Example 2.19: it exhibits a bounded function living in $\mathcal{L}[a, b]\backslash\mathcal{R}[a, b]$.) For the space $\mathcal{L}[a, b]$, the **functional** defined by Lebesgue integral is an **extension** of the Riemann integral, because for $f \in \mathcal{R}[a, b]$ the value for $\int_a^b f(x)dx$ gets independent of which integral from these two is being considered.

Watch out, when dealing with the so-called **improper** Riemann integrals, which means – allow us to recall – either $[a, b]$ fails to be finite or the taken functions lack to be bounded, such a pattern is altered, as shown in

[11] We would prefer the term **bounded** rather than finite, as the latter may push the idea of a set with a finite number of points.

Example 2.8 The bounded function $f(x) := (1/x)\sin x$ is not Lebesgue integrable[12] on the line $\mathbb{R}$, but its improper integral in the sense of Riemann exists. In fact, integration by parts gives

$$\lim_{B\to\infty}\int_{\pi/2}^{B}\frac{\sin x}{x}dx = -\lim_{B\to\infty}\int_{\pi/2}^{B}\frac{\cos x}{x^2}dx$$

and the right-hand side limit exists.

Analogously, the bounded function

$$f(x) := \begin{bmatrix} x^{-1}\sin(1/x) & 0 < x \le 1 \\ 0 & x = 0 \end{bmatrix}$$

fails to be Lebesgue integrable, since $\int_{1/n}^{1}|f(x)|dx \uparrow \infty$. Nevertheless, as

$$\int_{1/n}^{1}\frac{1}{x}\sin(1/x)dx = \int_{1}^{n}\frac{\sin(x)}{x}dx,$$

its improper integral in Riemann sense exists.

On this whole section, the characteristic function of any set A is denoted by

$$\Psi_A(x) := \begin{bmatrix} 1 & x \in A \\ 0 & x \notin A \end{bmatrix}. \tag{2.35}$$

Example 2.9 The function $f(x) := \Psi_{\mathbb{Q}}(x)$, characteristic of the rational numbers, fails to be Riemann integrable at no matter which interval $[a, b]$ is chosen. Indeed, for any partition

$$\Delta := \{a = x_0 < x_1 < \ldots < x_n = b\},$$

we obtain for the upper sums

$$\overline{S_\Delta}(f) := \sum_{\imath=1}^{n}(x_\imath - x_{\imath-1})\left[\sup_{x_{\imath-1}\le x\le x_\imath} f(x)\right] = b - a,$$

while the lower sums

$$\underline{S_\Delta}(f) := \sum_{\imath=1}^{n}(x_\imath - x_{\imath-1})\left[\inf_{x_{\imath-1}\le x\le x_\imath} f(x)\right] = 0,$$

[12] Confirm this claim with help from the results at Sect. 2.14.6.

and thus the upper integral

$$\overline{\int_a^b} f(x)dx := \inf_{\Delta} \overline{S_\Delta}(f)$$

equals $(b-a)$, being thus unequal to the lower integral

$$\underline{\int_a^b} f(x)dx := \sup_{\Delta} \underline{S_\Delta}(f) = 0.$$

It is worth remarking that, no matter the way the rationals are counted, say via $\{q_n\}$, by defining

$$f_N(x) := \begin{bmatrix} 1 & x = q_\iota 1 \le \iota \le N \\ 0 & \text{elsewhere} \end{bmatrix},$$

it follows that

$$f_N \in \mathcal{R}[a,b], \lim_{N\to\infty} f_N(x) = \Psi_{\mathbb{Q}}(x).$$

This indicates that the (monotonous) pointwise limit f of a sequence of functions from $\mathcal{R}[a,b]$ may fail to be Riemann integrable, even if $\|f\|_\infty < \infty$. (Compare this example with the statement of the Monotonous Convergence Theorem, Sect. 2.14.6.)

In Sect. 2.14.3, it is proved that $\Psi_{\mathbb{Q}}$ is Lebesgue-integrable. Moreover, its integral is zero, because the set of points where it does not vanish turns out to be, within this framework, "irrelevant."

Being $g := \Psi_{\mathbb{Q}} - (1/2)$, observe that g fails to be Riemann-integrable, but $|g| \in \mathcal{R}[a,b]$. (Such an event would never occur for Lebesgue-integral: arbitrary functions f and $|f|$ either are both – or both fail to be – Lebesgue-integrable.)

In the sense of Riemann, the lower integral of a function $f(x)$ may as well be defined as

$$\underline{\int_a^b} f(x)dx = \sup\{\int_a^b g(x)dx; g \in \mathcal{E}\},$$

with $\mathcal{E}$ denoting the set of all **step functions** defined on $[a,b]$:

$$g \in \mathcal{E} \Longleftrightarrow g := \sum_{\iota=1}^{n} c_\iota \Psi_{[x_{\iota-1}, x_\iota]},$$

for some mesh on $[a,b]$ and a set $\{c_\iota\}$ of constants. The functions g in $\mathcal{E}$ have their integrals **defined** by

$$\int_a^b g(x)dx := \sum_{\imath=1}^{n} c_\imath (x_{\imath-1}, x_\imath),$$

which is compatible with the notion of area under a curve graph, and from $\mathcal{E}$ we build the extension of the notion of integral for more general functions.
An alternate way to define Lebesgue integral is to get hold of $\mathcal{S}$, the set of the so-called **simple functions**, which means functions whose image is a finite set. These may be expressed in the form

$$f := \sum_{\imath=1}^{n} c_\imath \Psi_{E_\imath}$$

where the sets $E_\imath$ are spared to be intervals; rather, they ought to allow having a measure $m(E_\imath)$ assigned to each of them. In a precise fashion, we put

$$\int_a^b f(x)dx = \sum_{\imath=1}^{n} c_\imath m(E_\imath).$$

Being $\mathcal{S} \supset \mathcal{E}$, we expect then to be able to **approximate** a larger set of functions for which it will be possible to define their integrals.
When constructing the Riemann integral, we have looked at intervals in the x-axis, when we employed the step functions. For Lebesgue integral, emphasis is put on the intervals on the y-axis, as long as we now deal with the following **simple functions**: for $A \leq f(x) \leq B$, consider a partition $\{y_\imath\}$ for $[A, B]$, and then take $g \in \mathcal{S}$ with

$$g(x) := \sum_{\imath=1}^{n} y_\imath \Psi_{c_\imath},$$

where $c_\imath := f^{-1}([y_{\imath-1}, y_\imath])$. Observe that, because $c_\imath$ is not necessarily an interval, we need to impose conditions on f to be able to **measure** the sets $c_\imath$.
These facts stay unclear from the approach we have taken to build Lebesgue integral. In spite of that, these are ideas that turn out to be essential to justify the reason why we reach an integral much more powerful than Riemann's.

2.14.2 *Definition, Properties, and the Space $L^1(\mathbb{R})$*

Let us consider now, and in the sequel, $C_0(\mathbb{R})$ – the space of the real continuous functions defined on the line and that vanish outside a finite interval (which changes with each function). The expression

$$\|f\|_1 := \int |f(x)|dx$$

defines[13] a norm on $C_0(\mathbb{R})$. By making use of the functions from Exercise 2.17, it is observed that the normed space thus obtained is not complete. Denote its **completion** by $L^1(\mathbb{R})$. The elements in $L^1(\mathbb{R})$ are generally termed as **generalized functions**.

The main reason for such a nickname is the vanishing of our chances to mention – even to think of – any value at $x_0 \in \mathbb{R}$ for whatever $f \in L^1(\mathbb{R})$ is chosen. This is a straightforward consequence of the non-continuity of the functional

$$\begin{aligned} \delta_{x_0} : C_0(\mathbb{R}) &\to \mathbb{R} \\ f &\to \delta_{x_0}(f) := f(x_0) \end{aligned}$$

with respect to the norm $\|\cdot\|_1$. Just remind the discussion of choices to be made when we treated the PCE – Principle of Continuous Extension.

Despite that, it remains quite surprising a counterpart: we will be able to deal with notions and operators to be apparently introduced in a **pointwise** way, *i.e.*, they require (or, rather, should require) knowledge of the value for f at all points in its domain.

At the present stage, with all shortcuts we have gone through, we are able to define the main actor at this section. As long as

$$L(f) := \int f(x)dx$$

is linear and continuous in $C_0(\mathbb{R})$, it possesses a unique linear continuous extension to the space $L^1(\mathbb{R})$. This extension is throughout called the **Lebesgue integral**, which consequently is defined for **generalized functions**.

All linear properties of the Riemann integral

$$\begin{aligned} \int f+g &= \int f + \int g \\ \int \alpha f &= \alpha \int f \end{aligned} \tag{2.36}$$

are clearly preserved because we have constructed a **linear extension**. Besides, for $f \in C_0(\mathbb{R})$, we have the **positiveness**,

$$f \geq 0 \Longrightarrow \int f(x)dx \geq 0, \tag{2.37}$$

a property which, at first view, should not be demanded from elements in $L^1(\mathbb{R})$. We would be pushed to this analysis since, in order to verify that $f \geq 0$, we should

[13] Here $\int$ indicates the integral on the whole line $\mathbb{R}$.

own the **pointwise** knowledge of f. This barrier will be bypassed with once more some help from the **PCE**.
Consider, for fixed $a \in \mathbb{R}$, the non-linear operator

$$\left. \begin{array}{rl} T_a : C_0(\mathbb{R}) & \to C_0(\mathbb{R}) \\ f & \to T_a f := f_a \end{array} \right],$$

where

$$f_a(x) := \left[\begin{array}{ll} f(x) & \text{if } f(x) \leq a \\ a & \text{if } f(x) > a \end{array} \right. .$$

This is the so-called **truncation operator**, and it fulfills

$$||T_a f - T_a g||_1 \leq ||f - g||_1.$$

Thus, T_a is uniformly continuous, and it is possible to have it continuously extended to the whole space $L^1(\mathbb{R})$. Observe then that for $f \in C_0(\mathbb{R})$,

$$f \leq a \Longleftrightarrow T_a f = f, \tag{2.38}$$

we can take (2.38) as a **definition**, when $f \in L^1(\mathbb{R})$.
By the same token, we define

$$f \geq a \Longleftrightarrow T_{-a}(-f) = -f.$$

The absolute value of a generalized function may be introduced then from the operator T_a, with the choice of $a := 0$, by means of:

$$|f| := f_+ + f_-,$$

where we denoted

$$f_- := -T_a(f),\ f_+ := -T_a(-f)(a := 0).$$

It follows from the definition that $|f| \in L^1(\mathbb{R})$ as well as that

$$\int |f| = \int f_+ + \int f_-. \tag{2.39}$$

Observe also that we have

$$\int f = \int f_+ - \int f_-. \tag{2.40}$$

In an alternate track to construct the Lebesgue integral, we define the integral for a function f, provided that (2.39) is finite, with the expression (2.40).

Exercise 2.38 Prove that, for $f, g \in L^1(\mathbb{R})$,

$$f, g \geq 0 \Longrightarrow f + g \geq 0.$$

Exercise 2.39 Prove that, for $\alpha > 0$, $f \in L^1(\mathbb{R})$,

$$|f| \leq \alpha \Longleftrightarrow f \leq \alpha \text{ and } f \geq -\alpha.$$

Exercise 2.40 Verify that, for $f \in L^1(\mathbb{R})$,

$$\lim_{N\to\infty} T_N f = f.$$

Exercise 2.41 Verify: if $f \in L^1(\mathbb{R})$,

$$f \geq 0 \Longleftrightarrow \exists f_n \in C_0(\mathbb{R}),\ f_n \geq 0 \text{and} f_n \to f.$$

Based on the result in Exercise 2.41, it becomes simple to verify that in $L^1(\mathbb{R})$ positiveness – property (2.37) – also holds for the Lebesgue integral.

2.14.3 Null Measure Sets

After reaching the domain of real numbers through the rationals completion trek, our track became to search a lighter characterization for the irrational numbers, and the choice pointed to decimal representation. Following this pattern, we will look now for an equivalent way to deeply "understand" the **generalized functions**. We will get convinced that they own also the right to be thought of as **functions**, with their well-known very meaning, as long as we are tolerant enough, letting them to stay undefined on sets from their domain allowed to be small enough.

A set $\mathcal{M} \subset \mathbb{R}$ is defined as having[14] **null measure** if it is possible to find, for any value of $\epsilon > 0$, a family of intervals $I_n := (a_n, a_n + \delta_n)$, $n = 1, 2, \ldots$, whose union $\cup_{n=1}^{\infty} I_n$ contains $\mathcal{M}$ and such that $\sum_n \delta_n < \epsilon$.

It is clear that the following are examples of null measure sets: countable sets, as well as **countable** unions of null measure sets. Observe that an **arbitrary** union of null measure sets may fail to have a null measure, just look at $\cup_{x=-\infty}^{\infty}\{x\}$. Discovery

[14] All over this chapter, both terms measure and integral will mean Lebesgue measure and integral.

of a non-countable null measure set is not a simple task: the most popular sample is the **Cantor set**, see [61], pp. 36 and 236.
When a particular property $\mathcal{P}$ is valid in the complement of a null measure set, it is said that $\mathcal{P}$ holds **almost everywhere**, shortened as *ae*. A function is then said to be "*ae* null" when the set of points from its domain where it does not vanish has null measure. Another example: the functions $f \equiv 0$ and $g \equiv \Psi_{\mathbb{Q}}$ – cf. (2.35) – coincide *ae*. Whenever we say that a sequence $f_n(x)$ converges *ae*, this means that pointwise convergence holds in a subset of the domain shared by all this sequence functions, being necessarily null this subset complement measure.
Example 2.18 in Sect. 2.4 describes a sequence $\{f_n\}$ which converges to the null function according to the norm $\|\cdot\|_1$, but which fails to converge at no matter which point in the domain is chosen. Now, this sequence contains subsequences that turn out to be pointwise convergent; $\{f_{2_k}\}$ is, namely, one of them. Such a state of affairs is general, as claimed by the

Theorem 2.3 (Riesz-Fischer) *Given in $C_0(\mathbb{R})$ any sequence $\{f_n\}$, which is known to be a Cauchy sequence as regards to the norm $\|\cdot\|_1$, it is always possible to draw from it an ae convergent subsequence $\{f_{n_k}\}$. (Allow the emphasis: except for a null measure set in its whole domain, such a subsequence converges pointwise.)*

This result points out to an easier way to "look at" the generalized functions from $L^1(\mathbb{R})$: they show themselves exactly as *ae* limits of Cauchy sequences in $C_0(\mathbb{R})$. Or else, described in a more explicit form:

Whenever two Cauchy sequences from $C_0(\mathbb{R})$ are chosen, both convergent to the same element in $L^1(\mathbb{R})$ (two **equivalent** sequences), Riesz-Fischer Theorem may be applied to exhibit two *ae* defined functions. It may be proven then that these functions *ae* coincide. This fact motivates thus the introduction of an equivalence relation within the set of functions f known to be the *ae* limit of Cauchy sequences in $C_0(\mathbb{R})$. A conclusion follows, namely: the elements in $L^1(\mathbb{R})$ are in 1–1 correspondence with these equivalence classes, and we are thus allowed to think on the **generalized functions** from $L^1(\mathbb{R})$ as **classes of functions *ae* defined.**

This just presented remark justifies to claim that $\Psi_{\mathbb{Q}} \in L^1(\mathbb{R})$. Further, it is worth stating **Riesz-Fischer Theorem** for functions from $L^1(\mathbb{R})$:

Assuming that $f_n \in L^1(\mathbb{R})$, $n = 1, 2, \ldots$, and $\|f_n - f\|_1 \to 0$, we conclude the existence of at least a subsequence $\{f_{n_k}\}$ which *ae* converges to f.

Put an eye on the proof of the result stated on Exercise 2.10(a) and have it compared with the one obtained for the exercise that follows, provided use is made of this just presented formulation of the Riesz-Fischer Theorem.

Exercise 2.42 Verify that $\{f \in L^1(\mathbb{R}) \mid |f| \leq 1\}$ is a closed subset of $L^1(\mathbb{R})$.

Exercise 2.43

(a) Show that $\Psi_{(a,b)} \in L^1(\mathbb{R}) \iff (a, b)$ is bounded.
(b) Verify: the piecewise continuous functions belong to $L^1(\mathbb{R})$.

Although having dealt in the present section with only functions on the real line, all concepts introduced remain valid for $\mathbb{R}^n, n \geq 2$. Moreover the construction is alike, in either case. Of course, even with its unexpected secrets and surprises, the line $\mathbb{R}$ is easier to treat than the spaces $\mathbb{R}^n, n \geq 2$, as regards to technical details throughout this topic. This is particularly valid when defining the measure – even null measure – for a set. In spite of mentioning all over the real line, we try to keep away from any reasoning which is private to it.

We define the **measure of an open set** A as

$$m(A) := \sup \left\{ \int_{\mathbb{R}} f(x)dx;\ f \in C_0(\mathbb{R}),\ f \leq \Psi_A \right\}.$$

It can be verified that, being A an interval (a, b), or n−polycube, which means $A := \Pi_{\iota=1}^{n}(a_\iota, b_\iota)$, the following expressions hold:

$$m(A) = b - a \text{ or } m(A) = \Pi_{\iota=1}^{n}(b_\iota - a_\iota).$$

One could think as natural to define

$$m(A) := \int \Psi_A(x)dx, \tag{2.41}$$

but we would be stuck for not (yet) being aware whether $\Psi_A \in L^1(\mathbb{R})$. Such is the motivation for the notion of **measurable** sets, which are those that can receive a measure, cf. Sect. 2.14.7.

Exercise 2.44 On the real line, any open set Q may be described as the disjoint union of open intervals $Q_\iota = (a_\iota, b_\iota)$:

$$Q = \cup_{\iota=1}^{N} Q_\iota,\ N \leq \infty,\ Q_\iota \cap Q_\jmath = \emptyset,\ \iota \neq \jmath.$$

Verify that

$$m(Q) = \sum_{\iota=1}^{N}(b_\iota - a_\iota).$$

Exercise 2.45 Enumerate the rational numbers as $\{q_\iota\}$ and take $\epsilon > 0$. Consider the open set

$$Q_\epsilon := \cup_{\iota=1}^{\infty}(q_\iota - \epsilon/2^\iota, q_\iota + \epsilon/2^\iota).$$

This is a dense open set on the line. Verify: its measure is $\leq 2\epsilon$.

As long as we have in mind the notion of a measure tailored for sets more general than intervals, we are allowed to exchange the previous definition by the one which follows:

A set $\mathcal{N}$ is said to have **null measure** if, for any $\epsilon > 0$, there exists an open set $A_\epsilon \supset \mathcal{N}$ such that

$$m(A_\epsilon) \leq \epsilon.$$

2.14.4 The Spaces $L^p(\mathbb{R})$, $1 < p < \infty$

Now choose $1 < p < \infty$ and denote by $L^p(\mathbb{R})$ the completion of $C_0(\mathbb{R})$ when equipped with the norm $||\cdot||_p$, defined by

$$||f||_p := \left(\int_{\mathbb{R}} |f(x)|^p dx\right)^{1/p}.$$

It is easily seen that in $C_0(\mathbb{R})$, the functional $||\cdot||_p$ is non-negative, is homogeneous, and vanishes only for $f \equiv 0$. As regards to the triangle inequality, which has earned in this framework the name of Minkowski inequality, it will be shown with the help of another inequality, which owns a lot of importance, by itself:

Hölder's Inequality If $f, g \in C_0(\mathbb{R})$, it is seen that

$$\int |fg| \leq ||f||_p ||g||_q \tag{2.42}$$

as long as $p, q \in (1, \infty)$ are **conjugated exponents**, *i.e.*, it holds for them

$$\frac{1}{p} + \frac{1}{q} = 1 \tag{2.43}$$

or, the equivalent identities

$$pq = p + q, \tag{2.43'}$$

$$p = q/(q-1) = 1/(q-1) + 1, \tag{2.43''}$$

$$q = p/(p-1) = 1/(p-1) + 1. \tag{2.43'''}$$

Hölder's inequality (2.42) is a consequence of the relation

$$ab \leq \frac{a^p}{p} + \frac{b^q}{q}, \tag{2.44}$$

which is valid for $a, b \geq 0$ and conjugated exponents p, q. Besides, this inequality generalizes the well-known relationship between the arithmetic and geometric averages.
As seen in the figure exhibited below, the rectangle area is always smaller than the sum of the areas $A_1 + A_2$. Moreover, since

$$A_1 = a^p/p, A_2 = \int_a^b y^{1/(p-1)} dy = b^q/q,$$

(2.44) follows.

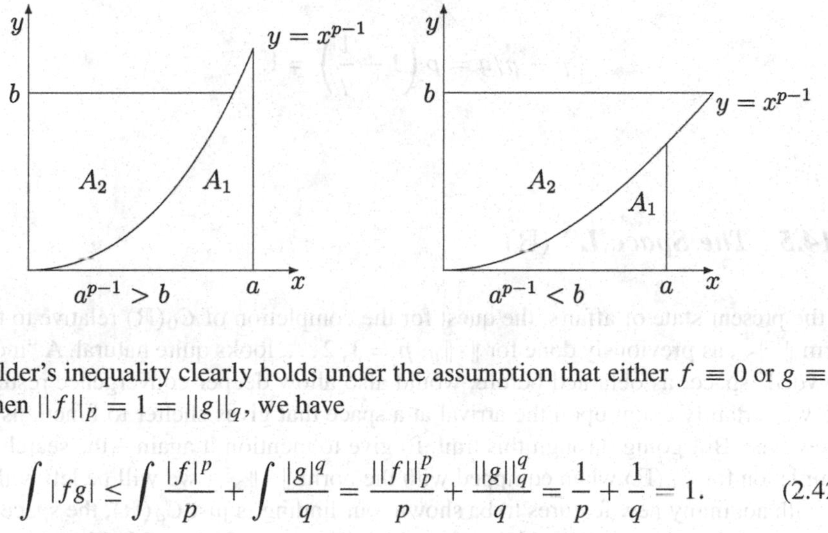

Hölder's inequality clearly holds under the assumption that either $f \equiv 0$ or $g \equiv 0$. When $||f||_p = 1 = ||g||_q$, we have

$$\int |fg| \leq \int \frac{|f|^p}{p} + \int \frac{|g|^q}{q} = \frac{||f||_p^p}{p} + \frac{||g||_q^q}{q} = \frac{1}{p} + \frac{1}{q} = 1. \tag{2.42$'$}$$

Finally, for arbitrary and non-null f, g, we can apply (2.42′) to $f/||f||_p$ and $g/||g||_q$ from which (2.42) follows.

Exercise 2.46 Deduce Hölder's inequality for the norms $|| \cdot ||_p$ introduced in $\mathbb{R}^N$ or in ℓ_0^∞ by

$$||x||_p := \left(\sum_{j=1}^n |x_j|^p\right)^{1/p}, \begin{bmatrix} N \in \mathbb{N} \text{ for } \mathbb{R}^N \\ N = \infty \text{ for } \ell_0^\infty \end{bmatrix}.$$

Minkowski's Inequality Whenever $f, g \in C_0(\mathbb{R})$, we obtain the so-called Minkowski's inequality:

$$||f+g||_p \leq ||f||_p + ||g||_p. \tag{2.45}$$

Since

$$|f+g|^p \leq |f+g|^{p-1}(|f|+|g|)$$

and, due to Hölder's inequality,

$$\int |f||f+g|^{p-1} \leq ||f||_p \left(\int |f+g|^{(p-1)q}\right)^{1/q} = ||f||_p \cdot ||f+g||_p^{p/q},$$

it follows, when $f+g \neq 0$, that

$$||f+g||_p^p \leq (||f||_p + ||g||_p)||f+g||_p^{p/q}.$$

From this relation, (2.45) may be deduced because

$$p - p/q = p\left(1 - \frac{1}{q}\right) = 1.$$

2.14.5 *The Space* $L^\infty(\mathbb{R})$

At the present state of affairs, the quest for the completion of $C_0(\mathbb{R})$ relative to the norm $\|\cdot\|_\infty$, as previously done for $\|\cdot\|_p$, $p = 1, 2 \ldots$, looks quite natural. A "more crowded" space, as obtained before, would also allow deeper convergence results, and we certainly count upon the arrival at a space that gives shelter to functions of a new type. But going through this trail, forgive to mention it again – the search of completion for $C_0(\mathbb{R})$ when equipped with the norm $\|\cdot\|_\infty$ – we will be left with a set with not many new features to be shown: our finding is just $C_a(\mathbb{R})$, the space of **continuous** functions null at ∞, *i.e.*, for which $\lim_{|x|\to\infty} f(x) = 0$ holds.
Had we have chosen a larger set, with less restrictions, namely, $C_L(\mathbb{R})$, the space of continuous and bounded real functions defined on the whole line, with the norm $||\cdot||_\infty$, its completion would not take us much more abroad: the starting point would already have been a complete normed space.
Thus, as long as we are looking for **more** functions, particularly for *ae* defined ones, as well as measuring them with the help of the norm $||\cdot||_\infty$ – indeed, another norm which generalizes it – an alternate trail must be followed.

Let f be a **measurable** function, which means: f is the *ae* limit of functions from $L^1(\mathbb{R})$. For $\beta > 0$, denote

$$I_\beta := [-\beta, \beta]^c = (-\infty, -\beta) \cup (\beta, \infty).$$

Define then

$$||f||_\infty := \inf\{\beta;\ f^{-1}(I_\beta)\text{has null measure}\}, \tag{2.46}$$

with the convention that $||f||_\infty := +\infty$ if the set for which we consider the "inf" is empty. Now define

$$L^\infty(\mathbb{R}) := \{f\text{measurable};\ ||f||_\infty < \infty\}.$$

The right-hand side of (2.46) is called the **essential supremum** of f. Observe that f may be modified on a null measure set in a such a way as to hold the following identity:

$$\sup_x |f(x)| = \sup_x \text{ess}\,|f(x)|.$$

Several applications make use of the claim stated below.

Theorem $L^\infty(\mathbb{R})$ *is a Banach space.*

As a matter of fact, let $\{f_n\}$ be a Cauchy sequence. Introduce

$$A_k := \{x;\ |f_k(x)| > ||f_k||_\infty\}$$

and

$$B_{m,n} := \{x;\ |f_m(x) - f_n(x)| > ||f_m - f_n||\},$$

then denote by E the following null measure set:

$$E := \cup_{k,m,n=1}^\infty A_k \cup B_{m,n}.$$

We change $f_k(x)$; in E, by making $f_k(x) = 0$, and then conclude that $\{f_n\}$ is a Cauchy sequence of bounded functions. We can claim then its (uniform) convergence to a function $f \in L^\infty(\mathbb{R})$, on the complement of E.

We will adopt the tradition of considering $p := 1$ and $q := \infty$ as conjugate exponents, since Hölder's inequality holds for these values: being $g \in L^1(\mathbb{R})$, $f \in L^\infty(\mathbb{R})$, we have

$$\int |fg| \leq ||f||_\infty ||g||_1.$$

2.14.6 Convergence Theorems

Section 2.14.3 introduces an equivalence relation for functions that share the property of being pointwise *ae* limit of Cauchy sequences in $C_0(\mathbb{R})$. For these sequences, it was *a priori* known that **convergence in norm** holds and therefore

$$\lim \int f_n = \int \lim f_n.$$

Next step is to pose the following question for a sequence $f_n \in L^1(\mathbb{R})$ which is known to satisfy

$$f_n(x) \xrightarrow{n} f(x) ae.$$

Is it possible to assure that:

(i) $f \in L^1(\mathbb{R})$?
(ii) $\int f_n \xrightarrow{n} \int f$?

The examples below show that neither one necessarily holds.

Example 2.10 Choose $f \geq 0$, $f \neq 0$, $f \in C_0(\mathbb{R})$, and $f_n(x) := f(x-n)$. It is seen that $f_n \to 0$ pointwise, but

$$\int f = \lim_n \int f_n > 0 = \int \lim_n f_n. \tag{2.47}$$

Example 2.11 For $f_n := \Psi_{[0,n]}/n$, (2.47) also holds, being even **uniform** the convergence of the sequence f_n.

In both examples above, the sequence of functions f_n turns out to be uniformly bounded, both in pointwise sense and with respect to the norm $\|\cdot\|_1$. Observe, on the other hand, that there exists no function $g \in L^1(\mathbb{R})$ such that

$$|f_n| \leq g \forall n.$$

A sequence which satisfies such a property is said to be **dominated** by g. Further, the following important result holds:

Dominated Convergence Theorem (Lebesgue) *Whenever a given sequence f_n converges to $f ae$ and further is dominated by some function $g \in L^1(\mathbb{R})$, it then follows that*

(i) $f \in L^1(\mathbb{R})$
(ii) $\|f_n - f\|_1 \xrightarrow{n} 0$.

Consequently:

$$\lim_n \int f_n = \int f = \int \lim_n f_n.$$

Exercise 2.47

(*a*) Calculate

$$\lim_{n\to\infty} \int_0^n \left(1 + \frac{x}{n}\right)^n \exp^{-2x} dx.$$

(*b*) Prove that, if $f(x) := |x|^\alpha \Psi_{(0,1]}(x)$, then

$$f \in L^1(\mathbb{R}) \Longleftrightarrow \alpha > -1.$$

This exercise illustrates the technique of function **truncating** coupled to the Dominated Convergence Theorem so as to calculate the value of integrals on unbounded sets.

The result which follows, besides its importance *per se*, allows us to refer to the Lebesgue integral for a function $f \geq 0$, or $f \leq 0$, even when $f \notin L^1(\mathbb{R})$.

Monotone Convergence Theorem (Lebesgue) *Suppose that, for* $n = 1, 2, \ldots$, $f_n \in L^1(\mathbb{R})$ *and*

$$0 \leq f_1 \leq f_2 \leq \ldots \leq f_n \uparrow f ae. \tag{2.48}$$

It follows, then:

$$f \in L^1(\mathbb{R}) \Longleftrightarrow \int f_n < M$$

and further,

$$\int f = \lim_n \int f_n. \tag{2.49}$$

Proof If $f \in L^1(\mathbb{R})$, by the Dominated Convergence Theorem, (2.49) holds. Thus, the integrals $\int f_n$ turn out to be uniformly bounded.

Conversely, if $\int f_n < M$, $\{f_n\}$ is a Cauchy sequence in $L^1(\mathbb{R})$, because, for $m > n$,

$$\|f_m - f_n\|_1 = \int |f_m - f_n| = \int (f_m - f_n) = \int f_m - \int f_n.$$

Further, being the sequence of real numbers $\{\int f_n\}$ monotone and bounded, it converges; hence, it is a Cauchy sequence. Then, from Riesz-Fischer theorem (see Sect. 2.14.3), it follows that $\lim f_n = f$ (in $L^1(\mathbb{R})$), so that, by making use of the Dominated Convergence Theorem, we reach (2.49).
Now let $f \geq 0$ and suppose that there exists a sequence of functions in $L^1(\mathbb{R})$ that satisfy (2.48). Define then the integral of f as the extended real[15] given by

$$\int f := \lim_n \int f_n.$$

It becomes then natural, for $f \leq 0$ and $0 \geq f_n \geq f_{n+1} \downarrow f$, to extend

$$\int f := -\int (-f).$$

Exercise 2.48 Prove: the above definitions are coherent – in mathematical slang, "good definitions" – *i.e.*, they do not depend on the chosen sequences; for $f \in L^1(\mathbb{R})$, it coincides with that one previously introduced.
We ought to observe that the definition above requires *a priori* the existence of function sequences that **approximate** f. It is worth to emphasize that, for arbitrary f, it is not always possible to find an approximating sequence, depending on the searching space as well as the chosen distance to measure the approximation. Recall that, in order to define a function f to be **measurable**, we have required from f to be the *ae* limit of functions from $L^1(\mathbb{R})$.

It is worth to compare the above theorem with the

Example 2.12 Let $f_n := \Psi_{[n,\infty)}$. It may be seen that:

$$f_n \geq f_{n+1} \downarrow 0,$$

but

$$\infty = \lim_n \int f_n \neq \int \lim_n f_n = 0.$$

On the other hand, if $g_n := -f_n$,

$$g_n \leq g_{n+1} \uparrow 0,$$

but

[15] *I.e.*, an element of $\mathbb{R} \cup \{-\infty, \infty\}$.

$$\lim_n \int f_n \neq \int \lim_n f_n.$$

Remind that, for an arbitrary sequence of reals $\{\alpha_n\}$, the concept of lim inf, defined as

$$\liminf_{n\to\infty} \alpha_n := \lim_{k\to\infty} \{\inf_{n\geq k} \alpha_n\} = \sup_k \{\inf_{n\geq k} \alpha_n\},$$

is an element of the extended real line, $\mathbb{R} \cup \{-\infty, +\infty\}$, which exists independently of the properties of the numbers $\{\alpha_n\}$.

The Monotone Convergence Theorem has as one of its consequences the

Fatou Lemma *Let $f_n \geq 0$ be arbitrary measurable functions. The inequality that follows always holds:*

$$\int \left(\liminf_{n\to\infty} f_n\right) \leq \liminf_{n\to\infty} \left(\int f_n\right). \tag{2.50}$$

Observe that, on the left-hand side, we integrate a function which is pointwise defined with the help of the concept of lim inf. A proof for this Lemma may be read in [62], pp.22.
It is worth mentioning that the strict inequality may hold in (2.50), cf. Examples 2.11 and 2.12 in this section.

2.14.7 Fubini Theorem and Differentiation × Integration

The elements of the spaces $L^p(\mathbb{R})$, for $1 < p < \infty$, are also treated as generalized functions like those living in $L^1(\mathbb{R})$. They would look more friendly if thought of as measurable functions *ae* defined and subject to the constraint $|f|^p \in L^1(\mathbb{R})$. The reader must get aware that both Hölder and Minkowski inequalities hold in $L^p(\mathbb{R})$, due to a reasoning based on continuity.

Exercise 2.49 A set E is considered as **measurable** when Ψ_E is a measurable function. Given a function f defined on a measurable set E, we say that $f \in L^p(E)$ if the extension

$$\hat{f}(x) := \begin{bmatrix} 0 & x \notin E \\ f(x) & x \in E \end{bmatrix}$$

belongs to $L^p(\mathbb{R})$. Define on $L^p(\mathbb{R})$ the equivalence relation R_E, for a measurable set E, by

$$f R_E g \iff f\Psi_E = g\Psi_E ae.$$

Then define the space $L^p(E)$ as the quotient space $L^p(\mathbb{R})/R_E$. Verify that such a definition for $L^p(E)$ is equivalent to the previously assigned.
The result that follows explains a characterization for those functions which are integrable on the sense of both Lebesgue and Riemann, along a bounded interval.

Theorem 2.1 *Let $f \in L^1(a,b)$ be bounded. Then $f \in \mathcal{R}[a,b]$ if and only if the set of its discontinuity points is countable.*[16]

As previously remarked, cf. Sect. 2.14.3, up to this point we have restricted ourselves, just for the sake of simplifying the exposition, to the integration on subsets of the real line $\mathbb{R}$. Nevertheless, in what follows, we ought to deal with integrals on subsets of $\mathbb{R}^n$, for $n \geq 2$.
Suppose $\mathcal{Q} := [a,b] \times [\alpha, \beta] \subset \mathbb{R}^2$ to be a bounded rectangle where the continuous function $f : \mathcal{Q} \to \mathbb{R}$ is defined. It is known as an important fact that, in order to obtain the value of the **double** integral $\int\int_{\mathcal{Q}} f(s,\tau) ds d\tau$, we may try to get hold of any of the two expressions described below. Both describe a relationship between a two variables integration and two (sequential) only one variable integrations. In other words, the point to study is whether, for double integrals, it is possible to employ **repeated** integrals so as to get:

$$\int\int_{\mathcal{Q}} f(s,\tau) ds d\tau \left\} \begin{aligned} &= \int_a^b \left[\int_\alpha^\beta f(s,\tau) d\tau \right] ds \\ &= \int_\alpha^\beta \left[\int_a^b f(s,\tau) ds \right] d\tau \end{aligned} \right. . \tag{2.51}$$

This result contains a point which, despite its importance, is poorly underlined, remaining almost unnoticed. Emphasis is normally concentrated on the process of exchanging the limits, or changing the order of integration, as told by (2.51). But, when we have at our disposal only weaker hypotheses – say, being sure only that f is Riemann-integrable – in order to be allowed to consider (2.51), how to guarantee the integrability of the (one variable) functions

$$\left. \begin{aligned} \phi_\tau(s) &:= f(s,\tau) \forall \tau; \ \Phi(\tau) := \int_a^b \phi_\tau(s) ds \\ \lambda_s(\tau) &:= f(s,\tau) \forall s; \ \Lambda(s) := \int_\alpha^\beta \lambda_s(\tau) d\tau \end{aligned} \right] , \tag{2.52}$$

whose integrals show up in (2.51)?
These are the results described by
Fubini Theorem [17] *Let*

[16] A proof for this result may be found, say, on [61], pp.248.
[17] This statement is restricted to $\mathbb{R}^2$ and to rectangles, but it may be generalized in different ways, like by taking arbitrary measurable sets $\mathcal{Q} := E \times F$, with $E, F \subset \mathbb{R}$, as well as $E \subset \mathbb{R}^m$, $F \subset \mathbb{R}^n$ also measurable, for any values of $m, n \in \mathbb{N}$. Their proofs and still more general statements may be read in [62].

$$f : \mathcal{Q} := [a, b] \times [\alpha, \beta] \subset \mathbb{R}^2 \to \mathbb{R}$$

be a measurable function. As long as $f \in L^1(\mathcal{Q})$*, if use is made of the notation introduced in (2.52), the claims that follows hold,*

$$\begin{array}{l} \phi_\tau \in L^1(a, b) \ ae \ in \ (\alpha, \beta), \ \Phi \in L^1(\alpha, \beta) \\ \lambda_s \in L^1(\alpha, \beta) \ ae \ in \ (a, b), \ \Lambda \in L^1(a, b) \end{array}$$

and (2.51) is true.

Exercise 2.50 For $\imath, \jmath = 0, 1, 2, \ldots,$ denote by $Q_\imath^\jmath$ the squares $[\imath, \imath + 1] \times [\jmath, \jmath + 1] \subset \mathbb{R}^2$, by $\Psi_\imath^\jmath := \Psi_{Q_\imath^\jmath}$, and take

$$a_{\imath\jmath} := \begin{bmatrix} 1 & \jmath = \imath \\ (1/2^\imath) - 1 & \jmath = \imath + 1 \\ 0 & \text{otherwise} \end{bmatrix},$$

so as to let

$$\left. \begin{array}{lll} f : & \mathbb{R}^2 & \to \mathbb{R} \\ & (x, y) & \to f(x, y) := \sum_{\imath,\jmath=0}^{\infty} a_{\imath\jmath} \Psi_\imath^\jmath(x, y) \end{array} \right].$$

Verify that, for this function – which does not belong to $L^1(\mathbb{R}^2)$ – the repeated integrals associated to (2.51) lead to different values.

Exercise 2.51 Adapt the previous exercise, attacking now a function

$$f : [0, 1]^2 \to \mathbb{R}.$$

Fubini theorem points to conditions under which we are entitled to change the order to evaluate the repeated integrals. Recall that in Chap. 1 we already indicated the needed care to invert two limit processes. We shall analyse now how to *differentiate under the integral sign*, in other words, the interaction of two distinct limit processes.
To make it explicit, consider the operator $f \to g$ defined on a real function space,

$$g(s) := \int_I K(x, s) f(x) dx, \tag{2.53}$$

where I is an interval – no tears if it is unbounded – and, for each s, or at least outside a null measure set, $K(\cdot, s) f(\cdot) \in L^1(\mathbb{R})$. Questions asking to be posed: Will $g(s)$ be differentiable when the partial derivative $\partial K(x, s)/\partial s$ exists? In such a case, will it be true that

$$\frac{d}{ds}g(s) = \int_I \frac{\partial}{\partial s}K(x,s)f(x)dx? \tag{2.54}$$

And if that holds, which are the precise hypotheses to require?
Now, to prove (2.54) amounts to reach

$$\lim_{h\to 0}[\Delta_{s,h}\int_I K(x,s)f(x)dx] = \int_I \lim_{h\to 0}[\Delta_{s,h}K(x,s)]f(x)dx$$

for the finite difference operator

$$\Delta_{s,h}\phi(s) := [\phi(s+h) - \phi(s)]/h,\, h \in \mathbb{R}. \tag{2.55}$$

Being linear such relation, it may be written as

$$\lim_{h\to 0}\int_I [\Delta_{s,h}K(x,s)]f(x)dx = \int_I \lim_{h\to 0}[\Delta_{s,h}K(x,s)]f(x)dx.$$

In order to have the right-hand side of (2.54) defined, it must be valid that

$$f(\cdot)\frac{\partial}{\partial s}K(\cdot,s) \in L^1(\mathbb{R}).$$

It is seen then that this is the only hypothesis to be required. Just observe that, under such conditions, the Dominated Convergence Theorem will hold, cf. the previous subsection.
Now, as it occurs with whatever result that deals with differentiation of a function sequence, it is mandatory to handle direct hypotheses about the differentiability of the limit function. Properties on derivatives or differentiability shared by the sequence elements are not passed along to the limit function (compare this with the result about distributions on Theorem 4.2, Sect. 4.4).
To finish, observe that, being the derivative a **local** operator, in order to verify the theorem stated in the sequel, one may get restricted to subintervals in the domain of f, so as to reach the required estimates for the operators (2.55); see, for example, [37].

Theorem 2.2 (Integral Differentiation) *Let g be defined by (2.53) and*

$$f(\cdot),\, K(\cdot,s)f(\cdot),\, f(\cdot)\frac{\partial}{\partial s}K(\cdot,s) \in L^1(\mathbb{R}).$$

Then g is differentiable, and its derivative fulfills the relation in (2.54).

2.14.8 Classical Bibliography Remarks

The construction of Lebesgue integral we have described follows essentially that one in [46, 47]; [60] and [62] introduce the notion of a measure through an axiomatic road; [7] is quite clear and bears many examples; and [31, 58], and [61] exhibit elegant and fairly compact presentations, each one with a particular approach. It is also worth to browse throughout [42].

Chapter 3
Dual of a Normed Space

3.1 Introduction

This chapter gets you familiar with some properties of the **linear forms** on a normed space V, by what are meant linear operators

$$\ell : V \to \mathbb{R}.$$

We adopt the usual notation of V' for the vector space of all linear forms defined on V (the so-called V **algebraic dual**), while V^* will represent the vector space of all **continuous linear** forms, or **functionals**, on V, also referred as V **topological dual**. It is always valid that $V^* \subset V'$, and we have remarked (Exercise 2.32) that V^* is a Banach space, as long as it is equipped with the norm defined in (2.28), being V complete or not. (Let us emphasize that, whenever the notation V^* is used, we will be mentioning this vector space of linear continuous forms, under the operators norm!)

3.2 Linear Forms and Hyperplanes

For any vector space V, a subspace H is said to be a **hyperplane** whenever, given another subspace W for which $H \subset W \subset V$, either $H = W$ holds, or $W = V$. In other words, H is a **maximal** subspace.

C. A. de Moura, *Functional Analysis Tools for Practical Use in Sciences and Engineering*, https://doi.org/10.1007/978-3-031-10598-2_3

Here follows an alternate definition:

Exercise 3.1 The subspace $H \subset V$ is a hyperplane if there exists $w \in V \backslash H$ such that, for each $v \in V$,

$$v = \alpha w + h \tag{3.1}$$

is valid, with $\alpha = \alpha(v) \in \mathbb{R}$ and $h = h(v) \in H$. We can also tell that $H \cup \{w\}$ generates the whole space V. We can even deduce that the expansion (3.1) is unique, which allows writing $V = H \bigoplus [\omega]$.

Given $\ell \in V'$, we denote by $ker(\ell)$ the kernel, or the null space, of ℓ:

$$ker(\ell) := \{v \in V; \ell v = 0\}.$$

If ℓ is non-null, $ker(\ell)$ is a hyperplane: indeed, by taking $w \in V$, with $\ell w = 1$, and defining, for each $v \in V$,

$$\alpha_v := \ell v \qquad h_v := v - \alpha_v w,$$

we then have that $h_v \in ker(\ell)$ and $v = \alpha_v w + h_v$ holds.
Conversely, given a hyperplane H, it is possible to obtain a linear form whose null space is precisely H. It is enough to attribute an (arbitrary) value to ℓw with v in (3.1) and then employ the linearity property in that expression in order to define ℓ:

$$\ell v := \alpha(\ell w).$$

Besides, we have that

$$ker(\ell_1) = ker(\ell_2), \ell_1, \ell_2 \in V' \Longrightarrow \ell_1 = \alpha \ell_2, \text{ for some real } \alpha.$$

When $\ell \in V^*$, $ker(\ell)$ is **closed**. In fact, the following fact holds:

A linear functional ℓ is continuous if and only if $ker(\ell)$ is closed.

The converse assertion may be deduced from another claim: if $\ell \in V'$ is unbounded, the image by ℓ of any ball fills up the whole line $\mathbb{R}$. Stare at the proof.
Consider arbitrary reals $\delta > 0$ and $\alpha > 0$. Since ℓ is unbounded, there exists $v = v_\delta$ with

$$\|v_\delta\| < \delta, \quad \ell v_\delta > \alpha.$$

Linearity implies then:

$$\ell\{t v_\delta, -1 \le t \le 1\} = [-\ell v_\delta, \ell v_\delta] \supset [-\alpha, \alpha].$$

Fix then a real $\beta \neq 0$ and take the set $L_\beta := \{v \in V; \ell v = \beta\}$.

Having ℓ a closed kernel, L_β is equally closed, as this set is just a translation for $ker(\ell)$:

$$L_\beta = \beta w + ker(\ell), \text{with} \ell w = 1.$$

As long as $0 \notin L_\beta$, it is possible to catch a ball $B(0;\delta)$ for which

$$B(0;\delta) \cap L_\beta = \emptyset.$$

But this claim contradicts the assumed claim about the image of $B(0;\delta)$ by the functional ℓ : it ought to fill up the whole real line $\mathbb{R}$.
Observe that the closure of a vector subspace is also a subspace, so that:

Being a maximal subspace, a hyperplane either is dense or closed.

This conclusion lets to characterize the linear forms from V' and V^* by:

$$\left.\begin{array}{ll} \ell \in V^* & \Longleftrightarrow ker(\ell) \text{is closed} \\ \ell \in V', \ell \notin V^* & \Longleftrightarrow ker(\ell) \text{is dense} \end{array}\right] .$$

Exercise 3.2

(i) Verify that, if $\phi \in C_0^\infty(\mathbb{R})$, then
$\phi = \psi'$, for some $\psi \in C_0^\infty(\mathbb{R}) \Longleftrightarrow \int_{\mathbb{R}} \phi(x)dx = 0$.

(ii) Prove that, for every $\phi \in C_0^\infty(\mathbb{R})$, there exists a representation

$$\phi = \theta' + \alpha\psi_0,$$

being $\psi_0 \in C_0^\infty(\mathbb{R})$ fixed and $\theta \in C_0^\infty(\mathbb{R})$ depending on ϕ.

Consider now this problem: given a linear continuous form ℓ on a hyperplane $H \subset V$, how to reach for ℓ an extension $\tilde{\ell} \in V^*$? The simplest cases, where $\ell \equiv 0$ or $\overline{H} = V$, have already been discussed. Now, being H a closed hyperplane, use the notation from (3.1) to conclude that any linear extension of ℓ must satisfy

$$\tilde{\ell}v = \alpha\tilde{\ell}w + \ell h.$$

Therefore, the only value left to be found is $\tilde{\ell}w$. Having that in mind, as long as V is a Banach space, the choice for this value does not matter as regards to continuity of $\tilde{\ell}$. And this shows up to be true because

$$\begin{aligned} |\tilde{\ell}v| \le |\alpha||\tilde{\ell}w| + \|\ell\|\|h\| &\le \max\{|\tilde{\ell}w|, \|\ell\|\}(|\alpha| + \|h\|) \\ &\le \max\{|\tilde{\ell}w|, \|\ell\|\}\beta\|v\| \end{aligned}.$$

The last inequality comes from being

$$|||v||| := \|h\| + |\alpha|$$

a norm equivalent to the norm employed on V, as can be deduced from Theorem 3.1, Sect. 2.13, since

$$\|v\| = \|\alpha w + h\| \leq \max\{\|w\|, 1\}|||v|||.$$

With these steps we have ended the proof of

Theorem 3.1 *Let H be a closed hyperplane of a Banach space V and let ℓ be a continuous linear functional on H. For any choice of $w \in V \backslash H$ and $\gamma \in \mathbb{R}$, we can claim continuity for the linear functional $\tilde{\ell} \in V'$ defined by*

$$\tilde{\ell} w := \gamma, \tilde{\ell} h := \ell h, h \in H.$$

This result should be compared with another extension theorem, to be discussed on Chap. 5, where we try to duly emphasize its importance. Here we are just praising Hahn-Banach theorem.

3.3 Riesz Representation Theorem

For a Hilbert space V, fix a vector $w \in V$. The functional

$$\begin{aligned} f_w : V &\to \mathbb{R} \\ v &\to f_w v := (v|w) \end{aligned} \tag{3.2}$$

is an element of V^*, as remarked on Example 2.12c, Sect. 2.5. Indeed,

$$\|f_w\| = \|w\|, \tag{3.3}$$

holds, since $f_w w = \|w\|^2$ and, thanks to Schwarz inequality, $\|f_w\| \leq \|w\|$. From being linear the inner product it follows the linearity of the transform

$$\left. \begin{aligned} J : V &\to V^* \\ w &\to J_w := f_w \end{aligned} \right] .$$

From (3.3) we see that J is an isometry, thus, 1-to-1. The functionals on $J(V)$ own consequently a **natural representation** relatively to the vectors in V.
It is just expected to listen the question: Will J be onto? In full words, given $f \in V^*$, is it always possible to determine a vector $w \in V$ so that $f = f_w$?

As long as such an element w exists, it is necessarily unique and orthogonal to $ker(f)$. Conversely, if $ker(f)$ is a (closed) hyperplane for which it is possible to find an orthogonal vector $\tilde{w} \neq 0$, $\tilde{w} \perp ker(f)$, then $f \in J(V)$.
The question posed above gets then a positive answer as long as it is possible to find, for each closed hyperplane H, a non-null vector $\tilde{w} \perp H$.
Suppose that $u \notin H$. This vector is not necessarily orthogonal to H, but from it we construct $w \in V$, $w \neq 0$, $w \perp H$.
Following an inspiration from $\mathbb{R}^3$, we search the vector w from the **orthogonal projection** of u on H: we determine another vector $h_0 \in H$ which minimizes (on H) the function $\|u - h\|$.
Use the fact that H is closed and $u \notin H$ to be allowed to denote

$$d: \text{ distance } (u, H) := \in f_{h \in H} \|u - h\| > 0.$$

Let $\{h_n\}$ be a sequence on H for which $\|u - h_n\| \to d$. We claim that this is a Cauchy sequence. Indeed, just employ parallelogram rule (2.11) in order to reach:

$$\begin{aligned} \|h_m - h_n\|^2 &= \|(h_m - u) - (h_n - u)\|^2 \\ &= 2\|h_m - u\|^2 + 2\|h_n - u\|^2 - \left\|2\left(\tfrac{h_m + h_n}{2} - u\right)\right\|^2 \\ &\leq 2\{\|h_m - u\|^2 + \|h_n - u\|^2 - 2d^2\} \to 0 \end{aligned}$$

if $m, n \to \infty$. Such inequality occurs due to being $(h_m + h_n)/2 \in H$. Since V is complete and H is closed, H is necessarily complete, and thus h_n converges to some vector $h_0 \in H$.
Therefore, by taking $w := u - h_0$, one concludes that $d = \|w\|$. Besides, $(w|h) = 0$, $\forall h \in H$ holds, then. In fact, take $h \in H$ arbitrary; for any $\alpha \in \mathbb{R}$, we have the following inequality

$$d^2 \leq \|v - \alpha h\|^2 = d^2 - 2(w|h)\alpha + \|h\|^2\alpha^2,$$

and this implies to be null the coefficient of α, $(w|h)$.
To finish, observe that if there exists another vector $h_1 \in H$, $h_1 \neq h_0$, for which $d = \|u - h_1\|$ holds, then $(h_0 + h_1)/2 \in H$, and, again due to (2.11), it follows that

$$\begin{aligned} \|u - (h_0 + h_1)/2\|^2 &= 2\|(u - h_0)/2\|^2 \\ &\quad + 2\|(u - h_1)/2\|^2 - \|(h_0 - h_1)/2\|^2 < d^2. \end{aligned}$$

But this contradicts being h_0 a minimizer for $\|u - h\|$, $h \in H$.
In short, we are able to announce the

Theorem (Riesz Representation) *Let H be a Hilbert space. To each continuous linear form $f \in H^*$, it corresponds a unique vector $w = w_f \in H$ for which*

$$fv = (w|v), \forall v \in H. \tag{3.2'}$$

Conversely, for each $w \in H$ fixed, (3.2) *defines a linear form $f = f_w$ which turns out to be an element of V^*, and for which:*

$$\|f_w\|_{H_*} = \|w\|_H. \tag{3.3'}$$

Exercise 3.3 Verify that, in any Hilbert space,

$$\|v\| = \sup_{\|w\|=1} (v|w).$$

3.3.1 Lax–Milgram Representation Theorem

Given a Hilbert space H, Riesz theorem allows to identify the continuous linear forms from H^* to the vectors in H. The current section presents a theorem that generalizes such a result.

Take B as a **bilinear** form on H, which means that to each pair (x, y) of elements from H, B associates a real number, in such a way that, for each chosen x, $B(x, \cdot)$ is a linear functional and, analogously, for any fixed y, $B(\cdot, y)$ is linear. Suppose, further, to be B **bounded**, by which we mean that there exists a constant C for which the following inequality holds:

$$|B(x, y)| \leq C\|x\|\|y\|, \forall x, y \in H.$$

Such a property implies that the functionals $\beta_x := B(x, \cdot)$, *i.e.*,

$$\begin{aligned} \beta_x : H &\to \mathbb{R} \\ y &\to \beta_x = B(x, y) \end{aligned}$$

are all bounded, as long as their norm satisfy $\leq C\|x\|$.
This structure poses quite automatically the question: What are the functionals on H^* that may be represented in terms of B in the above fashion? In different terms, which is the portion of H^* which can be generated from the linear forms defined throughout the expression that introduced β_x?
Observe: by Riesz theorem, given $x \in H$, it is possible to claim the existence of a unique $v = v(x) \in H$ such that

$$B(x, y) = (v \mid y), \forall y \in H. \tag{3.4}$$

So, it all amounts to determine the image of the bounded transformation

$$\begin{aligned} T_B : H &\to H \\ x &\to T_B x := v \end{aligned}$$

where v satisfies (3.4). We then impose the condition of being B **coercive**, by what is meant the existence of a constant $c > 0$ for which

$$|B(x, y)| \geq c\|x\|^2 \forall x \in H.$$

This way T_B is necessarily 1–1. Moreover, it turns out to be continuous the mapping $T_B^{-1} : V := Im(T_B) \to H$. To check this out, it suffices to make use of the Exercise 3.4, Chap. 2 as well the inequality

$$c\|x\|^2 \leq |B(x, x)| = |(T_B x|x)| \leq \|T_B x\| \|x\|.$$

As long as T_B and its inverse are continuous, the subspace V is necessarily complete, thus a Hilbert space. It is then possible to apply Riesz representation theorem to it, with respect to the functionals from V^*

$$v \to (v|u) \text{ with } u \in H \text{ arbitrary,}$$

and this gives

$$(v|u) = (v|w),$$

for some $w = w(u) \in V$ and every $v \in V$. Thus, $u - w \in V^{\perp}$. But

$$c\|u - w\|^2 \leq |B(u - w|u - w)| = (T_B\{u - w\}|u - w) = 0,$$

which conducts to

$$u - w = 0 \Longrightarrow u \in V \Longrightarrow V = H.$$

In short, these steps demonstrate the

Lax–Milgram Lemma.[1] Let B be a bilinear bounded and coercive bilinear form on the Hilbert space H, equipped with the inner product $(\cdot|\cdot)$. Then, to each vector $x \in H$, it is associated a unique vector $v = v(x) \in H$ for which

$$(x \mid u) = B(v, u), \forall u \in H.$$

It is worth to call the reader's attention: there is no need to assume B to be **symmetric**. It is plain clear the symmetry that lives on the link between the variables x and y on the reasoning we have described.

[1] This is the common way this representation theorem is mentioned in the literature.

Exercise 3.4 Verify that, under the hypotheses above on B, both

$$\|T_B\| = C, \|T_B^{-1}\| = 1/c$$

hold, provided

$$C := \sup_{\{\|x\|=\|y\|=1\}} B(x, y), c := \inf_{\{\|x\|=1\}} B(x, x).$$

Exercise 3.5 It follows from the triangle inequality,

$$|B(x, y) - B(\xi, \eta)| \leq |B(x, y) - B(x, \eta)| + |B(x, \eta) - B(\xi, \eta)| \leq C\{\|x - \xi\| + \|y - \eta\|\},$$

that, as a function from H^2 to $\mathbb{R}$, the form B, being bounded and bilinear, is continuous provided we take on H^2 any one of the norms

$$\|(u, v)\|_{p,H} := \|(\|u\|_H, \|v\|_H)\|_p, p = 1, 2, \infty.$$

3.3.2 *An Application: Stokes Equation*

In the sequel we present an application for the Lax–Milgram lemma, namely, a discussion on the stationary Stokes equation. It models the behavior of the velocity $u = u(x) \in \mathbb{R}^3$ and the internal pressure $p = p(x) \in \mathbb{R}$ for a fluid restricted to a region $\Omega \subset \mathbb{R}^3$:

$$\begin{aligned} &a) \quad && \left.\begin{aligned} -\nu\Delta u + \nabla p &= f \\ \nabla \cdot u &= 0 \end{aligned}\right] \text{ on } \Omega \\ &b) && \\ &c) && u = 0 \quad \text{on } \partial\Omega := \text{boundary of } \Omega. \end{aligned} \tag{3.5}$$

Here, $\nu > 0$ is a given constant (viscosity coefficient), and the external force f is previously determined.

We operate in an informal way, admitting all needed level of **regularity**[2] for u, p, and f.

Let us multiply (3.5a) by a function v, equally regular and which fulfills the same conditions (3.5b) and (3.5c) imposed on u. Then integrate on all of Ω. After an integration by parts, we obtain

$$\nu(\nabla u, \nabla v) - (p, \nabla \cdot v) = (f, v)$$

[2] See note after (2.27).

where, it should be observed, we denote by $(\cdot, \cdot)$ the inner product for the vector functions ($\Omega \to \mathbb{R}^3$), as well as for real functions ($\Omega \to \mathbb{R}$). The condition (3.5c) implies the vanishing of the boundary term in the first integration by parts, while (3.5b) conducts to deducing that the second term on the left-hand side is null also. Conclusion: u is the solution for the variational equation

$$\nu(\nabla u, \nabla v) = (f, v) \forall v \in V, \tag{3.6}$$

where V is the space of regular functions that satisfy (3.5b) and (3.5c).
Just as in Example 2.6 in Sect. 2.13, it is seen that the bilinear form $(\nabla u, \nabla v)$ is coercive:

$$(\nabla u, \nabla u) \geq \alpha \|u\|^2$$

with respect to the norm

$$\|u\|^2 := \int (u, u) dx.$$

Thanks to this fact, Lax–Milgram lemma assures the existence of $u \in V$ which fulfills (3.6).
The last step amounts to prove that, being u regular, it satisfies (3.5a).

3.4 The Projection Theorem

While studying the **linear** forms on a normed space V, it was essential to observe the correspondence that exists between those forms and the hyperplanes from V. Besides, the **continuous** linear forms are associated to the **closed** hyperplanes. On the other hand, in Riesz representation theorem, we have considered, for a given $\ell \in V^*$, the **orthogonal projection** on the closed hyperplane $ker(\ell)$. This section aims to study the orthogonal projection operators on arbitrary closed subspaces on a Hilbert space.
The main goal is to reach the proof of the projection theorem. We remark that this result, at the same time, generalizes Riesz representation theorem – RPT – and its demonstration depends on RPT. It will be established a 1-1 correspondence between the closed subspaces of a Hilbert space and the projection operators.
A linear operator $P : V \to V$ on a vector space is said to be a **projection** if $P^2 = P$. For these operators, by denoting the image of P by

$$Im(P) := \{v \in V; v = Pw for w \in V\},$$

we have that

$$V = Im(P) \bigoplus ker(P).$$

This means that V is the **direct sum** of $Im(P)$ with $ker(P)$, *i.e.*, for each $v \in V$, it is possible to determine a unique pair (v_1, v_2) such that

$$v_1 \in Im(P), v_2 ker(P),$$
$$v = v_1 + v_2.$$

Conversely, if $V = V_1 \bigoplus V_2$, we can find a projection operator P such that

$$V_1 = Im(P), V_2 = ker(P).$$

When V is a normed space, we suppose **bounded** the considered projections. In such cases, being P a projection operator, the space V is written as a direct sum of two **closed** subspaces. The corresponding converse also holds, provided that V be **complete**:

> Given a Banach space B and two closed subspaces F_1 and F_2 such that $B = F_1 \bigoplus F_2$, there exists a (unique) projection operator P for which $Im(P) = F_1$, $ker(P) = F_2$.

The proof for this result, based in one of the three basic principles of functional analysis, the **Closed Graph Theorem**, may be read in [67], pp. 237.
When dwelling in an arbitrary Banach space, it is not necessarily possible to guarantee the existence of projection operators. On the opposite side, on any Hilbert space, we are aware of a "large number" of such projections, a fact deduced from the

Projection Theorem *Let V be a Hilbert space and U be one of its closed subspaces. Then*

$$V = U \bigoplus U^{\perp},$$

with the notation $U^{\perp} := \{v \in V; (v|u) = 0, \forall u \in U\}$.

If $v = v_1 + v_2$, $v_1 \in U$, $v_2 \in U^{\perp}$, being this representation determined in a unique way, the operator

$$P_U : V \to V$$
$$v \to P_U v := v_1$$

is a projection operator. The continuity of P_U follows from Pythagoras theorem:[3]

$$\|v\|^2 = \|v_1\|^2 + \|v_2\|^2 \Longrightarrow \|P_U\| \leq 1.$$

[3] In fact, $\|P_U\| = 1$, since in U, $P_U \equiv$ identity.

This way, we have deduced that to each closed subspace of a Hilbert space, it corresponds a **projection operator**.

Proof Consider, for $v \in U$, fixed, the functional

$$\left.\begin{array}{ll}\ell_v : & U \quad \to \mathbb{R} \\ & v \in U \to \ell_v u := (u|v)\end{array}\right].$$

Since $\ell_v \in U^*$, Riesz representation theorem guarantees the existence of a unique $w = w(v) \in U$ for which

$$\ell_v u = (w|u), \forall u \in U.$$

Thus:

$$(w - v|u) = 0, \forall u \in U,$$

which gives us

$$w - v \in U^{\perp}.$$

Observe that

$$\text{dist}(v, U) = \|v - w\|$$

because, for $u \in U$,

$$\begin{aligned}\|v - u\|^2 &= \|w + (v - w) - u\|^2 = \|(v - w) + (w - u)\|^2 \\ &= \|v - w\|^2 + \|w - u\|^2 + 2(v - w|w - u) \\ &= \|v - w\|^2 + \|w - u\|^2 \geq \|v - w\|^2.\end{aligned}$$

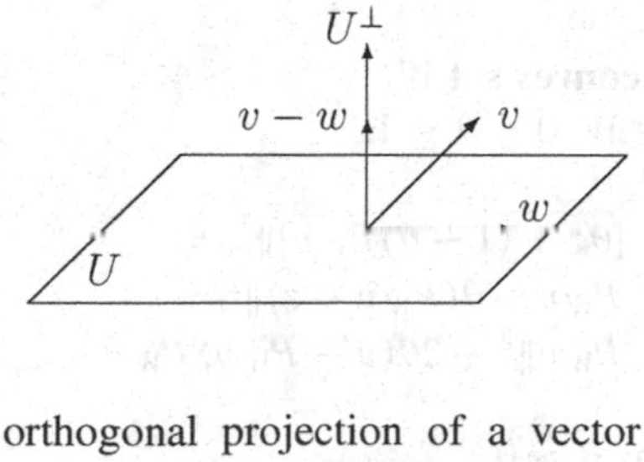

Geometrically, the orthogonal projection of a vector v on a closed vector subspace W is the vector w from W which gives the **best approximation** for v. In other words, which minimizes the function

$$f(z) := \|v - z\|, z \in W.$$

Therefore, we can infer that w is the solution of an optimization problem. Such a vector w is unique, and this result stays valid within a more general context, in which it is only required from W to be a closed convex[4] set.
This is stated in the

(Upgraded) Projection Theorem. *Let V be a Hilbert space and let $W \neq \emptyset$ be one of its closed convex subsets. Given any $v \in V$, there exists a unique $w = w(v) = P_W v$ in W for which*

$$\|v - w\| \leq \|v - z\|, \forall z \in W, \tag{3.7}$$

i.e.,

$$\text{dist}\,(v, W) = \inf_{z \in W} \|v - z\| = \|v - w\|. \tag{3.7'}$$

It is not hard to get convinced that neither of the stated hypotheses can be called off: if W is not closed, existence may fail to hold; and for non-convex W, more than one solution may show up. (It suffices to pick on $\mathbb{R}^2$,

$$v = (0, 0) \text{ and } W := \{(x, y); x > 1\},$$

in the first case, or

$$W := \{(x, y); x \geq 1 - y \text{ or } x \leq y - 1\},$$

in the second.) The theorem proof copies the steps of the one presented for Riesz theorem: construction of a minimizing sequence and self-convincing that its limit exists in W.
We say that the above defined operator

$$P_W : V \to W$$

is the **projection on the convex set** W.
Observe that, for any $z \in W, 0 \leq \theta \leq 1$,

$$\begin{aligned}\|v - P_W v\|^2 &\leq \|v - [\theta z + (1 - \theta) P_W v]\|^2 = \\ &= \|v - P_W v + \theta(P_W v - z)\|^2 = \\ &= \|v - P_W v\|^2 + 2\theta(v - P_W v, P_W v - z) + \theta^2 \|P_W v - z\|^2 .\end{aligned}$$

As a consequence, for any $\theta > 0$,

[4] A set C in a vector space is defined as convex if, given two elements $v_1, v_2 \in C$, the line segment which joins them stays within C. In alternate saying:

$$v_1, v_2 \in C \Longrightarrow t v_1 + (1 - t) v_2 \in C \forall t \in [0, 1].$$

$$2(v - P_W v, P_W v - z) + \theta \| P_W v - z \|^2 \geq 0,$$

and this lets to conclude that

$$(v - P_W v, P_W v - z) \geq 0, \forall z \in W. \tag{3.7''}$$

Conversely, if $w \in W$ satisfies

$$(v - w, w - z) \geq 0, \forall z \in W,$$

it follows that $w = P_W v$.
In fact,

$$\begin{aligned} \|v - z\|^2 &= \|v - w + w - z\|^2 \\ &= \|v - w\|^2 + 2(v - w, w - z) + \|w - z\|^2 \geq \|v - w\|^2. \end{aligned}$$

We can then conclude that (3.7), (3.7′), and (3.7″) are all equivalent formulations. The latter is called **variational inequality**. Geometrically, it expresses the condition that the vectors $v - P_W v$ and $z - P_W v$ define an angle $\theta > 90°$.
When W is a subspace, we can choose $z = y - P_W v$ in (3.7″), with arbitrary $y \in W$, from which it follows
$(v - P_W v, y) \geq 0, \forall y \in W$. Since we have $-y \in W$, the orthogonality condition is then deduced
$(v - P_W v, P_W v - z) = 0, \forall z \in W$.

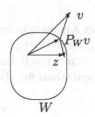

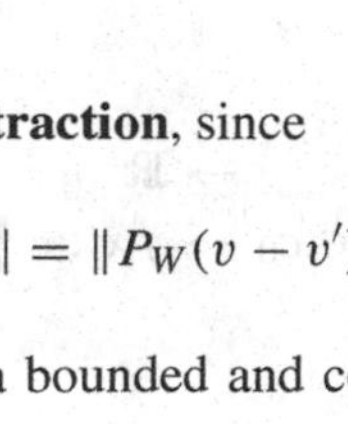

In this situation, $P_W v$ is a **contraction**, since

$$\|P_W v - P_W v'\| = \|P_W (v - v')\| \leq \|v - v'\|.$$

This property remains valid for a bounded and convex W. Indeed, it is seen to be true that

$$(P_W v' - v', z' - P_W v') \geq 0, \forall z' \in W. \tag{3.8}$$

By choosing in both (3.8) and (3.7″), respectively

$$z' := P_W v \text{ and } z := P_W v',$$

plus adding the resulting inequalities, we reach

$$(P_W v - v - P_W v' + v', P_W v' - P_W v) \geq 0$$

or

$$\begin{aligned}\|P_W v' - P_W v\|^2 &\leq (P_W v' - P_W v, v' - v)\\ &\leq \|P_W v' - P_W v\| \cdot \|v' - v\|.\end{aligned}$$

Exercise 3.6 Demonstrate:

(*a*) If N is a finite dimension subspace in a normed space, N is then closed.
(*b*) Let H be a Hilbert space and $F \subset H$ a finite dimension subspace. By the projection theorem, $H = F \bigoplus F^{\perp}$. For this particular (and important) case, prove directly such result, but without using Riesz theorem.

Example 3.1 The operators T_N introduced in Sect. 2.8 are projection operators. From the linear forms in Example 2.12a, Sect. 2.5, we can as well introduce projections in $\mathbb{R}^N$.

3.5 Representation for Some Dual Spaces

Riesz representation theorem shows a more concrete form to operate with bounded functionals on arbitrary Hilbert spaces. Nevertheless, for a given normed space, it may be impossible to get hold of a similar result. In spite of that, for some particular examples, some convenient **representations** may become available.

Example 3.2 The dual of c_0

Let $y = (y_j) \in \ell_1$. By defining

$$\left.\begin{array}{rl} f: \quad c_0 & \rightarrow \mathbb{R} \\ x = (x_j) \in c_0 & \rightarrow fx := \sum_{j=1}^{\infty} x_j y_j \end{array}\right], \tag{3.9}$$

the inequality

$$|fx| \leq \sum_{j=1}^{\infty} |x_j||y_j| \leq \|x\|_\infty \|y\|_1$$

indicates not only the convergence of the series described in (3.9) but also the boundedness for f. This way we are let to conclude:

$$f \in c_0^*.$$

Surprisingly, all functionals in c_0^* exhibit such a form:

Theorem 3.2 *Given $f \in c_0^*$, there exists $y = y_f = (y_j)_{j \in \mathbb{N}} \in \ell^1$ such that*

$$fx = \sum_{j=1}^{\infty} y_j x_j, \forall x = (x_j) \in c_0$$

and further

$$\|f\| = \|y\|_1. \tag{3.10}$$

Proof If such y exists, we must have $y_j = fe^j$, where

$$e^j = (e_i^j)_i = (\delta_i^j), \delta_i^j := \begin{bmatrix} 0 & i \neq j \\ 1 & i = j \end{bmatrix} \text{ (Kronecker delta).}$$

Let us define then $y_j := fe^j$ and try to verify if $y = (y_j)$ belongs to ℓ^1 and the announced representation holds.
Let

$$x^N := \sum_{j=1}^{N} \operatorname{sgn}(fe^j)e^j \in \ell_0^{\infty} \subset c_0,$$

$$\text{where} \begin{bmatrix} \operatorname{sgn}(a)a = |a|, \forall a \in \mathbb{R}, a \neq 0 \\ \operatorname{sgn}(0) = 0 \end{bmatrix}.$$

Thus we can deduce that

$$fx^N = \sum_{j=1}^{N} |fe^j| = \sum_{j=1}^{N} |y_j| \leq \|f\| \|x^N\|_{\infty} = \|f\|.$$

There we are let to claim that:

$$y \in \ell^1 \text{ and } \|y\|_1 \leq \|f\|.$$

As long as

$$\|\sum_{j=1}^{N} x_j e^j - x\|_\infty \overset{N}{\to} 0, \forall x = (x_j) \in c_0,$$

we get the conclusion that

$$fx = \sum_{j=1}^{\infty} y_j x_j$$

holds. From the inequality

$$|fx| \leq \sum_{j=1}^{\infty} |y_j x^j| \leq \|x\|_\infty \|y\|_1$$

it follows that $\|f\| \leq \|y\|_1$, which implies (3.10).

And from this result, we deduce the

Theorem 3.3 *Given $f \in c^*$, there exists $z = z_f \in \ell^1$, such that*

$$fx = z_1(\lim_j x_j) + \sum_{j=1}^{\infty} z_{j+1} x_j.$$

Proof Being c_0 a closed hyperplane from c, since

$$c = S \bigoplus c_0,$$

where $S := \{(\alpha, \alpha, \alpha, \ldots) \in \ell^\infty, \alpha \in \mathbb{R}\}$, from Theorem 3.1, Sect. 2.2 the above representation follows for c^*.

These identifications are known as **isometric isomorphisms**. This means that the operators

$$\begin{array}{c} T : c_0^* \to \ell^1 \\ f \to Tf := y_f \end{array} \quad \text{and} \quad \begin{array}{c} T : c^* \to \ell^1 \\ f \to Tf := y_f \end{array}$$

are linear, 1–1, onto and preserve the norm. Besides, we have the

Theorem 3.4 *Let $p, 1 \leq p < \infty$ and q, its **conjugated exponent**, given by*

$$q = \begin{bmatrix} q_p := p/(p-1) & p \neq 1 \\ q_1 := \infty & p = 1 \end{bmatrix}.$$

Then we have

$$(\ell^p)^* = \ell^q,$$

or, in a more precise form, the topological dual for ℓ^p may be **identified** *to the space ℓ^q:*

To each functional $f \in (\ell^p)^*$ it corresponds a unique element $y = y_f \in \ell^q$ such that

$$fx = \sum_{j=1}^{\infty} y_j x_j \forall x = (x_j) \in \ell^p,$$

and conversely. Besides, such correspondence is an **isometry**, *since*

$$\|f\| = \|y_f\|_q.$$

This result proof follows the reasoning line of the representation theorems above and is cordially left as an attractive chore to the reader.
Any one that gets aware of the previous result can't restrain from posing the question: Is it true that

$$(\ell^\infty)^* = \ell^1?$$

It is rather simple to verify that ℓ^1 may be identified to a **part** of $(\ell^\infty)^*$:

Exercise 3.7 For each $y \in \ell^1$,

$$\begin{aligned} f_y : \quad & \ell^\infty \quad \to \mathbb{R} \\ & x = (x_j) \to f_y x := \textstyle\sum_{j=1}^{\infty} x_j y_j \end{aligned}$$

is a functional in $(\ell^\infty)^*$ and

$$\|f_y\| = \|y\|_1.$$

It is possible to prove, although we do not have yet at our disposal the needed facts, that not every functional $f \in (\ell^\infty)^*$ bears the form f_y, for some element $y \subset \ell^1$.

We would like to point out that the – more intuitive – identification results hereby claimed for some sequence spaces aim at making the reader familiar with the concept of dual spaces representation. We start off from simpler frameworks for which the corresponding proofs are not tough to get.
All these results are special cases of another Riesz representation theorem which we formulate in a simpler form than the one commonly found, cf. [61].

Theorem (Riesz) *Given a measurable set $\Omega \subset \mathbb{R}^N$, let $1 < p, q < \infty$ be real conjugated exponents, and when $q := \infty$, take $p = 1$.*
To each functional $\ell \in [L^p(\Omega)]^*$, there corresponds a function

$$g = g_\ell \in L^q(\Omega),$$

unique, such that

$$\ell f = \int_\Omega f(x)g(x)dx \forall f \in L^p(\Omega), \tag{3.11}$$

and conversely. Moreover,

$$|||\ell||| = \|g_\ell\|_q. \tag{3.12}$$

Exercise 3.8 Prove that, for $g \in L^q(\Omega)$, the functional ℓ introduced by (3.11) belongs to $[L^P(\Omega)]^*$ and, in a direct fashion, that (3.12) holds.
It is important to observe that

$$L^1(\Omega) \overset{\subset}{\neq} [L^\infty(\Omega)]^*,$$

in the sense of such identification.
Some of the proofs hereby omitted, as well as some other results within this context, may be found in [6, 69], or [72].

3.6 The Bidual Space

For a normed space V, consider the topological dual for V^*, which gets the notation V^{**} and is called V **bidual**. We show at once several examples of elements from V^{**}, here are they:
Define the transformation

$$\left. \begin{array}{rl} J : V \to & V^{**} \\ v \to J_v : & V^* \quad \to \mathbb{R} \\ & \ell \in V^* \to J_v(\ell) := \ell v \end{array} \right] .$$

The linearity for J_v is quite evident, while its continuity follows from

$$|J_v(\ell)| = |\ell v| \le \|\ell\|_{V^*} \|v\|_V,$$

a relationship that implies

$$\|Jv\|_{V^{**}} \leq \|v\|_V \forall v \in V. \tag{3.13}$$

As a matter of fact, it will be proved on Chap. 4 that equality always holds in (3.13), which then implies that the mapping J is an **isometry** between V and a subspace of V^{**}. This mapping is usually mentioned as **canonical identification**.

Exercise 3.9

(*a*) Being $V := \ell^p$, or $V := L^p(\Omega)$, with $1 < p < \infty$, or else, if V is a Hilbert space, J is then onto.
(*b*) Wherever $\dim(V) < \infty$, then J is onto.

Those normed spaces for which the corresponding operator J is onto are said to be **reflexive**. Exercise 3.9 mentions the most important examples of reflexive normed spaces.
A strategy to verify that ℓ^∞ (or $L^\infty(\Omega)$) fails to be reflexive is to observe the existence of vectors $x, y \in \ell^\infty$ (or, respectively, $L^\infty(\Omega)$), with

$$\|x\|_\infty = \|y\|_\infty = \|x+y\|_\infty = \|x-y\|_\infty = 1.$$

But such a fact is forbidden to occur within reflexive spaces, as we can deduce from Millman's theorem, cf. [72], pp. 126.
A final remark is worth to listen to. Since V^{**} is complete and

$$\|Jv\|_{V^*} = \|v\|_V,$$

we already know: if it happens that V fails to be complete, the function J can not be onto, *i.e.*, V is not reflexive. In alternate words,

$$V \text{ reflexive} \implies V \text{ Banach} .$$

3.7 Radon-Nikodym Representation

Let us get back to the expression (2.41)

$$m(A) := \in t_{\mathbb{R}} \Psi_A(x)dx = \in t_A 1dx$$

which defines **Lebesgue measure** for a measurable set A on the real line. If the real 1 is exchanged for another positive real, the value of the measure assigned to all measurable sets on the line will be changed. But that will occur in a homogeneous way, independently of their location on $\mathbb{R}$, in other words, preserving Lebesgue measure property of being **translation invariant**. It would amount to assign different **weights** to different locations had we defined, for measurable $\mu : \mathbb{R} \to \mathbb{R}^+$,

$$\lambda(A) :=\in t_A \mu(x)dx. \tag{3.14}$$

Being $\mathcal{M}$ the class[5] of measurable sets on the line, the function

$$\lambda : \mathcal{M} \to [0, \infty]$$

introduces a new measure, because

$$\lambda(\bigcup_{\iota=1}^{M} A_\iota) = \sum_{\iota=1}^{M} \lambda(A_\iota),\ M \in \mathbb{N} \cup \{\infty\},\ A_\iota \in \mathcal{M} \text{ disjoint sets.}$$

Besides, it holds that

$$m(A) = 0 \Longrightarrow \lambda(A) = 0, \forall \text{ measurable} A. \tag{3.15}$$

This relation between two measures is denoted by $\lambda << m$. It is then said that λ is **absolutely continuous** with respect to m.
Radon-Nikodym representation theorem assures – see [60, 62] – that, if a given measure λ satisfies (3.15) relatively to Lebesgue measure in $\mathbb{R}^n$, then it necessarily bears the form (3.14), now for some measurable $\mu : \mathbb{R}^n \to \mathbb{R}^+$ and any measurable $A \in \mathbb{R}^n$.
The function μ is then known as the **Radon-Nikodym derivative** for λ with respect to m, the Lebesgue measure in $\mathbb{R}^n$.

3.8 Dirac Terminology

While looking for a mathematical formalism to model physical phenomena at atomic particles level, Paul Dirac [29] has made use of the symmetry described in Riesz representation theorem. Borrowing the term *bracket* associated to the inner product symbol $< | >$, he has made the distinction between the vectors on the left of the bracket, which are the functionals in H^*, and the vectors on the right, *i.e.*, those that belong to the very space H. Then he would mention them, respectively, as *bra* and *cket* and employ the notation $< v|$ and $|w >$.
Both *bra* and *cket* are vectors from the space H, but the ones quoted with *bra* get to be identified with the elements of the dual space. When they get together (through the inner product), they originate a scalar, and the product by a scalar is denoted

---

[5] This class composes a $\sigma-$**algebra**, which means: (i) $\emptyset, \mathbb{R} \in \mathcal{M}$; (ii) $A \in \mathcal{M} \Longrightarrow A^c \in \mathcal{M}$ (iii) $A_\iota \in \mathcal{M}, M \in \mathbb{N} \cup \{\infty\} \Longrightarrow \cup_{\iota=1}^{M} A_\iota \in \mathcal{M}$.

by $\alpha|w>$ or $<v|\alpha$, depending on being employed a *cket* or a *bra*. By the same fashion, for the linear operators, either the notation $T|w>$ or $<v|T$ is employed. This way, to define an eigenvector, we write

$$T|w>=\lambda|w> \quad \text{or} \quad <v|T=<v|\lambda.$$

Given a mapping T and the vectors $<v|$ and $|w>$, the real numbers

$$\{<v|T\}|w> \quad \text{and} \quad <v|\{T|w>\}$$

are, in principle, distinct. If, for a particular operator, equality for these values is always true, *i.e.*, if associativity holds, such an operator is called **self-adjoint** – Dirac option was for the term **real**. For these operators it is allowed, no ambiguity showing up, to exclude the braces, writing them then as $<v|T|w>$.

Chapter 4
Sobolev Spaces and Distributions

4.1 Introduction and Notation

The present chapter aims to discuss the concepts of distributions and of Sobolev spaces, whose presence can not to be forgotten whenever one deals with many problems on differential equations – either from a theoretical or numerical viewpoint.

Most of the examples exhibited in the previous chapters restricted themselves to functions of just one real variable. We present now the comfortable Laurent Schwartz notation: it allows writing in a quite compact form expressions that deal with any kind of derivatives for arbitrary functions of several variables.

Let $n \geq 1$ be fixed. Once n non-negative integers $\imath_1, \imath_2, \ldots, \imath_n$ are chosen, together with an arbitrary point $\xi := (\xi_1, \xi_2, \ldots, \xi_n)$ on $I\!R^n$, we denote

$$\left.\begin{array}{r} \imath := (\imath_1, \imath_2, \ldots, \imath_n) \\ |\imath| := \imath_1 + \imath_2 + \ldots + \imath_n = \sum_{\jmath=1}^n \imath_\jmath \\ \imath! := \imath_1!\imath_2!\ldots\imath_n! = \Pi_{\jmath=1}^n \imath_\jmath! \\ \xi^\imath := \xi_1^{\imath_1}\xi_2^{\imath_2}\ldots\xi_n^{\imath_n} = \Pi_{\jmath=1}^n \xi_\jmath^{\imath_\jmath} \end{array}\right] .$$

As long as $\imath_\jmath$ are indices, we say that $\imath$ is a **multi-index** and, for example, $a_\imath$ will stand for $a_{\imath_1\imath_2\ldots\imath_n}$. For two multi-indices $\imath$ and ℓ, denote

$$\binom{\imath}{\ell} := \frac{\imath!}{\ell!(\imath-\ell)!} = \frac{\imath_1!\imath_2!\ldots\imath_n!}{\ell_1!\ell_2!\ldots\ell_n!(\imath_1-\ell_1)!\ldots(\imath_n-\ell_n)!} .$$

By using the symbol $D_\jmath$ for the partial differentiation operator $\partial/\partial x_\jmath$, with $1 \leq \jmath \leq n$, then D will denote the **gradient vector**

C. A. de Moura, *Functional Analysis Tools for Practical Use in Sciences and Engineering*, https://doi.org/10.1007/978-3-031-10598-2_4

$$D := (D_1, D_2, \ldots, D_n)$$

so that we use the representation

$$D^\iota := D_1^{\iota_1} D_2^{\iota_2} \ldots D_n^{\iota_n} = \frac{\partial^{|\iota|}}{\partial x_1^{\iota_1} \partial x_2^{\iota_2} \ldots \partial x_n^{\iota_n}}.$$

Throughout this chapter, the norms $\|\cdot\|_p$ in $I\!R^n$ shall change their notation to $|\cdot|_p$, while the Euclidean norm $|\cdot|_2$ will get still simpler: $|\cdot|$.

The exercises below illustrate how practical Schwartz notation is.

Exercise 4.1 (The Binomial Theorem) For $x, y \in I\!R^n$ and $\iota, \jmath, k$ multi-indices, it holds, for this finite sum:

$$(x + y)^\iota = \sum_{\jmath + k = \iota} \frac{\iota!}{\jmath! k!} x^\jmath y^k.$$

Exercise 4.2 If $x \in I\!R^n$, $|x|_\infty < 1$, the generalized geometric series gives

$$\sum_\iota x^\iota = \frac{1}{(1 - x_1)(1 - x_2) \ldots (1 - x_n)}.$$

Exercise 4.3 If $x \in I\!R^n$ and $|x|_1 < 1$, the identity below holds,

$$\sum_\iota \frac{|\iota!|}{\iota!} x^\iota = \frac{1}{(1 - x_1 - x_2 - \ldots - x_n)}.$$

Exercise 4.4 For $x \in I\!R^n$ and any integer $m > 0$, it is true that:

$$\sum_{|\iota| = m} \frac{m!}{\iota!} x^\iota = (x_1 + x_2 + \ldots + x_n)^m.$$

Exercise 4.5 (Leibnitz Formula) For $f, g : I\!R^n \to I\!R$ smooth enough,[1] we have

[1] As a matter of fact, it suffices to suppose, through almost all frameworks we deal with, that all acting functions have as their domains not the full $I\!R^N$ space but some of its open subsets. The needed additional hypotheses will be described therein.

$$D^\iota[f(x)g(x)] = \sum_{\jmath+k=\iota} \frac{\iota!}{\jmath!k!}[D^\jmath f(x)][D^k g(x)].$$

Exercise 4.6 When $f : I\!R^n \to I\!R$ has derivatives of all orders, its Maclaurin series is given by

$$f(x) = \sum_\iota \frac{1}{\iota!} D^\iota f(0) x^\iota.$$

4.2 Sobolev Spaces $H^k(\Omega)$ and $H_0^k(\Omega)$

Given an open subset $\Omega \subset I\!R^n$ and an integer $k \geq 0$, denote by $C^k(\Omega)$ the space of all real functions defined on Ω such that $D^r f$ is continuous for $|r| \leq k$. Following a previous convention, $C_0^k(\Omega)$ will represent the subspace of $C^k(\Omega)$ composed by all functions f null outside a compact $\subset \Omega$ – which may change with each considered f. (The term **compact** on $I\!R^n$ means a bounded and closed n–dimensional set.) To complete, let

$$C^\infty(\Omega) := \cap_{k=0}^\infty C^k(\Omega),\ C_0^\infty(\Omega) := \cap_{k=0}^\infty C_0^k(\Omega).$$

On $C_0^k(\Omega)$ we use the notation below for the norms thus defined,

$$\left[\begin{array}{l} \|f\|_{r,p} := \left(\sum_{|\iota|\leq r} \int_\Omega |D^\iota f(x)|^p dX\right)^{1/p},\ dX := dx_1 \ldots dx_n \\ \|f\|_{r,\infty} := [\sup_{x\in\Omega;|\iota|\leq r} |D^\iota f(x)|] \end{array}\right],$$

where $0 \leq r \leq k$. Naturally, for $r = 0$, $\|f\|_{0,p} = \|f\|_p$.

From now on, we will deal only with $\|\cdot\|_{k,2}$, for which we simply choose the notation $\|\cdot\|_k$, and further introduce

$$C_*^\infty(\Omega) := \{f \in C^k(\Omega); \|f\|_k < \infty\}.$$

Both spaces, $C_0^k(\Omega)$ and $C_*^k(\Omega)$, equipped with the norm $\|\cdot\|_k$, lack completeness. These spaces completion receive the notation, respectively, of $H_0^k(\Omega)$ and $H^k(\Omega)$. It is quite clear that they are Hilbert spaces and their elements are generalized functions reached from a procedure similar to that one which has led to $L^1(I\!R)$ (cf. Sect. 2.14).

For $k = 0$, it is seen that

$$H^k(\Omega) = H_0^0(\Omega) = L^2(\Omega). \tag{4.1}$$

On $C_0^k(\Omega)$ and $C_*^k(\Omega)$, the inner product is given by

$$(f|g)_k := \sum_{|\iota|\leq k} \left(D^\iota f | D^\iota g\right),$$

where

$$(f|g) := \int_\Omega fg dX$$

denotes the inner product on $L^2(\Omega)$.

Given $u \in H^k(\Omega)$, there exists $\{u_\jmath\}$ in $C_*^k(\Omega)$ for which $\|u_\jmath - u\|_k \to 0$. Since, for every multi-index α with $|\alpha| \leq k$, we have that

$$\|D^\alpha u_\jmath - D^\alpha u_\ell\|_0 \leq \|u_\jmath - u_\ell\|_k,$$

we conclude that $\{D^\alpha u_\jmath\}$ is a Cauchy sequence in $L^2(\Omega)$. As long as $L^2(\Omega)$ is complete, there exists $v_\alpha \in L^2(\Omega)$ such that $\|D^\alpha u_\jmath - v_\alpha\|_0 \to 0$. It is then possible to characterize $H^k(\Omega)$ as a set of functions from $L^2(\Omega)$ such that there exists $\{u_\jmath\}$ in $C_*^k(\Omega)$ and v_α in $L^2(\Omega)$ that fulfill

$$\lim_\jmath \|u_\jmath - u\|_0 = 0, \lim_\jmath \|D^\alpha u_\jmath - v_\alpha\|_0 = 0, |\alpha| \leq k. \tag{4.2}$$

Being $u \in C_*^k(\Omega)$, if $\|u_\jmath - u\|_k \to 0$, then (4.2) remains valid with $v_\alpha = D^\alpha u$. Such a result inspires to think on the functions v_α in (4.2) as a kind of generalization for the concept of derivative for functions in $H^k(\Omega)$. We make sharp such a generalization with the

Definition 4.1 Given a function $u \in L^2(\Omega)$, we say that the functions $v_\alpha \in L^2(\Omega)$, $|\alpha| \leq m$, are their **derivatives in the strong sense** if there exists a sequence $\{u_\jmath\}$ in $C_*^k(\Omega)$ such that

$$\lim_\jmath \int_\Omega |u_\jmath - u|^2 dX = \lim_\jmath \int_\Omega |D^\alpha u_\jmath - v_\alpha|^2 dX = 0, |\alpha| \leq m. \tag{4.2a}$$

This fact is also expressed by saying that the function u has all derivatives (in the strong sense) of order $\leq m$.

At once a problem pops up: What about the **uniqueness** for these derivatives? From the divergence theorem, the integration by parts formula follows: if $\overline{\Omega}_0 \subset \Omega$ and $\phi, \psi \in C_*^k(\Omega)$, it is deduced that

$$\int_{\Omega_0} \phi D_\iota \psi dX = \int_{\partial\Omega_0} \phi\psi\eta_\iota dS - \int_{\Omega_0} \psi D_\iota \phi dX, 1 \leq \iota \leq n \tag{4.3}$$

with $\eta := (\eta_1, \ldots, \eta_n)$ the exterior normal to Ω_0. Now, just take $\phi \in C_0^k(\Omega)$ on (4.3) to arrive at

$$\int_\Omega \phi D_\imath u_k dX = -\int_\Omega u_k D_\imath \phi dX, 1 \leq \imath \leq n$$

or, more generally,

$$\int_\Omega \phi D^\alpha u_k dX = (-1)^\alpha \int_\Omega u_k D^\alpha \phi dX, |\alpha| \leq k. \tag{4.4}$$

From (4.2) and (4.4), it may be claimed that

$$\int_\Omega \phi v_\alpha dX = (-1)^\alpha \int_\Omega u_k D^\alpha \phi dX, |\alpha| \leq k, \forall \phi \in C_0^k(\Omega). \tag{4.5}$$

If the functions w_α as well as v_α would be derivatives for u in the strong sense, we would necessarily get

$$\int_\Omega \phi v_\alpha dX = \int_\Omega \phi w_\alpha dX, \forall \phi \in C_0^k(\Omega),$$

or else,

$$(\phi | v_\alpha - w_\alpha) = 0, \forall C_0^k(\Omega).$$

But since $C_0^k(\Omega)$ is dense in $L^2(\Omega)$ – cf. (4.1) – the conclusion $v_\alpha = w_\alpha$ would follow.

For $k = 0$, $H^k(\Omega) = H_0^k(\Omega)$ holds. If $k \geq 1$, this result is false, in general. For example, let Ω be bounded and $k = 1$. For

$$z \in I\!R^n, |z| = 1 \text{ and } f(x) := \exp[< x|z >] = \exp \sum_{\jmath=1}^n x_\jmath z_\jmath,$$

it is seen that

$$D_\imath f(x) = z_\imath f(x), \Delta f(x) = |z|^2 f(x) = f(x).$$

Thus, for any $\phi \in C_0^1(\Omega)$,

$$\begin{aligned}(f|\phi_\imath)_1 &= (f|\phi) + \sum_{\jmath=1}^n (D_\jmath f | D_\jmath \phi) \\ &= (f|\phi) - (\Delta f|\phi) = 0.\end{aligned}$$

Now, $f \in C^\infty(I\!R^n)$, therefore $f \in C_*^k(\Omega) \subset H^1(\Omega)$ for all $k \geq 0$, since Ω is bounded. As long as $C_0^1(\Omega)$ is dense in $H_0^1(\Omega)$, we can not have $f \in H_0^1(\Omega)$, because $f \perp C_0^1(\Omega)$.

Conclusion $H_0^1(\Omega) \neq H^1(\Omega)$, for bounded Ω.

Take now $\Omega := I\!R^n$ so as to prove that $C_0^k(\Omega)$ is dense in $C_*^k(\Omega)$. Indeed, consider in $I\!R^n$ the function $\psi(x) := \theta(|x|)$, where θ was employed in Exercise 2.17, Sect. 2.7. Given $f \in C_*^k(\Omega)$, it may be verified that

$$f_\jmath(x) = \psi(x/\jmath)f(x) \to f(x)$$

with respect to the norm $\|\cdot\|_k$. (This is the so-called **truncation** of f.) Then

$$f_\jmath(x) - f(x) = 0 \text{if} |x| \leq \jmath,$$

and thus

$$\|f_\jmath - f\|_k^2 = \sum_{|\imath|\leq k} \int_{|x|\geq \jmath} |f(x) - D^\imath[\psi(x/\jmath)f(x)]|^2 dX.$$

But, according to Exercise 4.5, we have

$$\begin{aligned} D^\imath[\psi(x/\jmath)f(x)] &= \sum_{p+\ell=\imath} \frac{\imath!}{p!\ell!}[D^p f(x)][D^\ell \psi(x/\jmath)] \\ &= \sum_{p+\ell=\imath} \frac{\imath!}{p!\ell!} D^p f(x) \frac{1}{\jmath^{|\ell|}} D^\ell \psi(x)|_{x=x/\jmath}. \end{aligned}$$

Since $\psi \in C_0^\infty(I\!R^n)$, $\|D^\ell \psi\|_{0,\infty}$ is bounded for every ℓ, with $|\ell| \leq k$, and this implies that

$$\|f_\jmath - f\|_k^2 \leq C \sum_{|\imath|\leq k} \int_{|x|\geq \jmath} |D^p f(x)|^2 dX, \tag{4.6}$$

for some constant C.

The Dominated Convergence Theorem assures then that, as long as $\jmath \to \infty$, the right-hand side of (4.6) tends to 0, which closes the proof of

Theorem 4.1 *Whenever $k \geq 0$ is an integer, we have*

$$H^k(I\!R^n) = H_0^k(I\!R^n).$$

Exercise 4.7 If $0 \le k \le \ell$, then $H^k(\Omega) \supset H^\ell(\Omega)$ and

$$\begin{aligned} \imath : H^\ell(\Omega) &\to H^k(\Omega) \\ f \quad &\to \imath(f) := f \end{aligned}$$

is bounded, with norm ≤ 1. (An analogous result holds for all spaces $H_0^k(\Omega)$.)

Exercise 4.8 Define the mapping

$$D : H^k(\Omega) \to H^{k-|\alpha|}(\Omega)$$

for $|\alpha| \le k$ and prove its boundedness, with $\|D^\alpha\| \le 1$.

4.3 Weak Derivative and Regularization

Another notion of derivative lives in $H^k(\Omega)$, namely, the **derivative in the weak sense**. A function $v_\alpha \in L^2(\Omega)$ is said to be the derivative in the weak sense, of order α, for $u \in H^k(\Omega)$ if (4.5) holds for any $\phi \in C_0(\Omega)$. We can verify that this notion of derivative is equivalent to the one previously introduced, despite being operationally simpler.

Clearly, if v_α is the order α derivative of $u \in H^k(\Omega)$ in the strong sense, it is also its derivative in the weak sense. We prove now the reciprocal claim. The main importance of this proof is that it calls for the help from a strongly used technique, the **regularization**. It is essentially from [32] the following

Proof Let ϕ be the bell function on $I\!R$, introduced in Exercise 2.1, Chap. 2, and denote by ρ the $n-$dimensional bell function, *i.e.*, $\rho(x) := C\phi(|x|)$, where $C := [\int_{I\!R^n} \phi(|x|)dX]^{-1}$, and thus $\int_{I\!R^n} \rho dX = \int_{|x|\le 1} \rho(x)dX = 1$. Given a compact $K \subset \Omega$ and $\epsilon > 0$, with $\epsilon < \text{dist}(K, \partial\Omega)$, being $u, v_\alpha \in L^2(\Omega)$, if (4.5) holds, take

$$J_\epsilon v(y) := \epsilon^{-n} \int_\Omega \rho\left(\frac{y-x}{\epsilon}\right) v(x)dX, \forall v \in L^2(\Omega). \tag{4.7}$$

This function is called a **regularization** for v. It was due to Friedrichs the introduction of the operator J_ϵ. This strategy is based on a C^∞ function, and, since

$$\begin{aligned} J_\epsilon v(y) &= \epsilon^{-n} \int_{|y-x|\le\epsilon} \rho\left(\tfrac{y-x}{\epsilon}\right) v(x)dX \\ &= \int_{|z|\le 1} \rho(z)v(y-\epsilon z)dZ, \end{aligned} \tag{4.8}$$

it follows that

$$J_\epsilon v(y) - v(y) = \int_{|z|\le 1} \rho(z)[v(y-\epsilon z) - v(y)]dZ. \tag{4.9}$$

Making use of Schwarz inequality on (4.8), it is possible to conclude that

$$
\begin{aligned}
|J_\epsilon v(y)|^2 &= \left| \int_{|z|\le 1} \rho(z)^{1/2}[\rho(z)^{1/2} v(y-\epsilon z)]dZ \right|^2 \\
&\le \left\{ \left[\int_{|z|\le 1} \rho(z)dZ \right]^{1/2} \left[\int_{|z|\le 1} \rho(z)|v(y-\epsilon z)|^2 dZ \right]^{1/2} \right\}^2 \\
&= \left[\int_{|z|\le 13131} \rho(z)dZ \right] \int_{|z|\le 1} \rho(z)|v(y-\epsilon z)|^2 dZ \\
&= \int_{|z|\le 1} \rho(z)|v(y-\epsilon z)|^2 dZ,
\end{aligned}
$$

and, therefore,

$$
\int_K |J_\epsilon v(y)|^2 dY \le \int_{|z|\le 1} \left[\int_K |v(y-\epsilon z)|^2 dY \right] \rho(z) dZ,
$$

where we have been based on Fubini theorem. From that we can deduce that

$$
\|J_\epsilon v\|_{L^2(K)} \le \|v\|_{L^2(K_0)} \tag{4.10}
$$

for any compact $K_0 \subset \Omega$ whose interior contains K and for which it holds that dist $(K, \Omega \backslash K_0) > \epsilon$. Let now $\delta > 0$ be such that

$$
\int_{K_0} |v-w|^2 dX < \delta.
$$

From the linearity of J_ϵ and from (4.10), it is deduced that

$$
\int_{K_0} |J_\epsilon v - J_\epsilon w|^2 dX < \delta,
$$

while from (4.9) it follows that, if $\epsilon \to 0$, $J_\epsilon w(y) \to w(y)$ uniformly on K. The conclusion is then that, being $\epsilon > 0$ sufficiently small, we have

$$
\int_{K_0} |J_\epsilon w - w|^2 dX < \delta.
$$

These three last inequalities may be combined, so as to reach:

$$
\int_{K_0} |J_\epsilon v - v|^2 dX \to 0 \text{if} \epsilon \to 0, \forall v \in L^2(\Omega). \tag{4.11}
$$

We can then use the fact that the function $\rho\left(\frac{x-y}{\epsilon}\right)$ belongs to C_0^∞ as well as the definition of derivative in the weak sense so as to obtain, with use of (4.7) applied to $v := u$,

$$\begin{aligned} D^\alpha[J_\epsilon u(y)] &= \epsilon^{-n} \int_\Omega D_y^\alpha \rho\left(\tfrac{x-y}{\epsilon}\right) u(x) dX \\ &= \epsilon^{-n} \int_\Omega \rho\left(\tfrac{x-y}{\epsilon}\right) v_\alpha(x) dX, \end{aligned} \tag{4.12}$$

where the first inequality follows from the derivation under the integral sign, already discussed. It is quite clear that the right-hand side term in (4.12) equals to $J_\epsilon v_\alpha(y)$, that is, $D^\alpha(J_\epsilon u) = J_\epsilon v_\alpha$, and this implies that the (classical) differentiation of the regularization for u equals to the regularization of the derivative of u.

We then apply (4.11) to $v := v_\alpha$, in order to get

$$\int_{K_0} |J_\epsilon v_\alpha - v_\alpha|^2 dX \to 0 \text{if} \epsilon \to 0,$$

or, due to the above observation,

$$\int_{K_0} |D\alpha(J_\epsilon u) - v_\alpha|^2 dX \to 0. \tag{4.13}$$

Based on (4.11) and (4.13), it is deduced that the regularizations J_ϵ for u produce the sequence of functions u_j employed in the definition of the strong derivative of u, except for the following point: the integration on Ω in (4.2a) gets replaced by another one, where K turns out to be the domain of integration in (4.11) and (4.13).

In short, what remains proved is just the **local existence** of the order α derivative of u in the strong sense. The leap to the general framework is carried out by getting hold of the technique known as partition of unity, which we shall omit. The reader may read about it on, say, [32] or [2].

4.4 The Distributions

Throughout this section, Ω denotes an open connected[2] set from $I\!R^n$. As its first step, we introduce the space of the **test functions** $\mathcal{D}(\Omega)$. It is meant to be the set $C_0^\infty(\Omega)$ equipped with the notion of convergence below described.

Given $\phi_j \in C_0^\infty(\Omega)$, it is said that ϕ_j **converges to** ϕ **on** $\mathcal{D}(\Omega)$, denoted as $\phi_j \stackrel{\mathcal{D}}{\to} \phi$, if:

[2] An open set $A \subset I\!R^n$ is said to be connected if, for any pair of points p and $q \in A$, there exists a polygonal $P \subset A$ that joins p to q.

(*i*) There exists a compact $K \subset \Omega$ such that $\phi_J(x) = 0$ if $x \notin K$, for all $J = 1, 2, \ldots$.

(*ii*) For any multi-index α, with $0 \leq |\alpha|$, we have

$$\lim_{J\to\infty} \sup_{x\in\Omega} |D^\alpha \phi_J(x) - D^\alpha \phi(x)| = 0,$$

i.e., the functions ϕ_J and their derivatives converge uniformly (with respect to x) to the function ϕ and to all of its corresponding derivatives; or, even in other terms, if for each fixed k, we have that

$$\|\phi_J(x) - \phi(x)\|_{k,\infty} \to 0 \text{ if } J \to \infty.$$

It is then also verified that $\phi C_0^\infty(\Omega)$.

We have kept away from introducing a norm on $C_0^\infty(\Omega)$, being restricted to the notion of convergence. As it was remarked on Chap. 2, mostly all topological notions we will make use of may be defined having sequences as their basic support.

Example 4.1 Taking ρ as the bell function on $I\!R^n$, introduced on Sect. 4.3, it is verified that

$$\rho_n(x) := \rho(x/n)/n$$

converges uniformly towards zero. The same occurs for all its derivatives, but convergence in the sense of $\mathcal{D}(I\!R^n)$ does not hold. In fact, (i) can not be checked to occur.

It is straightforward to verify that, being $\phi, \phi_J, \psi, \psi_J \in \mathcal{D}(\Omega)$ and $\gamma_J, \gamma, \beta_J, \beta \in I\!R$, for $J = 1, 2, \ldots,$ it holds:

$$\left.\begin{array}{l} \phi_J \overset{\mathcal{D}}{\to} \phi,\ \gamma_J \to \gamma \\ \psi_J \overset{\mathcal{D}}{\to} \psi\ \ \beta_J \to \beta \end{array}\right] \Longrightarrow \gamma_J\phi_J + \beta_J\psi_J \overset{\mathcal{D}}{\to} \gamma\phi + \beta\psi, \tag{4.14}$$

$$\phi_J \overset{\mathcal{D}}{\to} \phi \Longrightarrow D^\alpha\phi_J \overset{\mathcal{D}}{\to} D^\alpha\phi, \forall\alpha \text{ multi-index.} \tag{4.15}$$

Motivated by the definition introduced on Sect. 2.5, in the present framework, we also define an operator T in $\mathcal{D}(\Omega)$ as **continuous** if $\phi_J \overset{\mathcal{D}}{\to} \phi$ implies $T\phi_J \to T\phi$.

Exercise 4.9 Verify that the following operators are continuous in $\mathcal{D}(\Omega)$:

$$
\begin{array}{lll}
(a) & \left.\begin{array}{l} D^\alpha : \mathcal{D}(\Omega) \to \mathcal{D}(\Omega) \\ \quad \phi \quad \to D^\alpha \phi \end{array}\right] & \alpha \text{ multi-index} \\
(b) & \left.\begin{array}{l} \delta_{\bar{x}} : \mathcal{D}(\Omega) \to I\!R \\ \quad \phi \quad \to \phi(\bar{x}) \end{array}\right] & \bar{x} \in \Omega \text{ fixed} \\
(c) & \left.\begin{array}{l} \delta_{\bar{x}}^\alpha : \mathcal{D}(\Omega) \to I\!R \\ \quad \phi \quad \to D^\alpha \phi(\bar{x}) \end{array}\right] & \bar{x} \in \Omega \text{ fixed}, \alpha \text{ multi-index} \\
(d) & \left.\begin{array}{l} M_\psi : \mathcal{D}(\Omega) \to \mathcal{D}(\Omega) \\ \quad \phi \quad \to \psi\phi \end{array}\right] C^\infty(\Omega) \text{ fixed} & \\
(e) & \left.\begin{array}{l} \mathcal{I} : \mathcal{D}(\Omega) \to I\!R \\ \quad \phi \quad \to \int_\Omega \phi dX \end{array}\right] &
\end{array}
$$

The continuous linear functionals[3] defined on $\mathcal{D}(\Omega)$ are named **distributions**. The distributions space is denoted by $\mathcal{D}'(\Omega)$, despite being more coherent with the notation used here to keep $\mathcal{D}^*(\Omega)$. For $T \in \mathcal{D}(\Omega)$, $\phi \in \mathcal{D}(\Omega)$, we will denote from now on

$$< T, \phi > := T(\phi).$$

Nevertheless, we will rest with $\mathcal{D}(a, b)$ instead of $\mathcal{D}((a, b))$ for open intervals on the line, *i.e.*, when $\Omega = (a, b)$; the same is applied to $\mathcal{D}'(a, b)$.

Example 4.2 All functionals on $(b), (c), (e)$ from Exercise 2.9 are seen to be distributions.

Example 4.3 Let f be a **locally integrable** function on Ω, which means, for each compact $K \subset \Omega$, it is assumed that $f \in L^1(K)$. (It is common to denote as $f \in L^1_{\text{loc}}(\Omega)$.)

Define the distribution $T(f)$ by

$$< T(f), \phi > := \int_\Omega f \phi dX, \forall \phi \in \mathcal{D}(\Omega). \tag{4.16}$$

Observe: we let $f \notin L^1(\Omega)$, but despite that the integral in (4.16) always exists, for every $\phi \in \mathcal{D}(\Omega)$. This results from the fact that the integral is only taken on a particular compact subset in Ω, outside which ϕ vanishes.

The linearity of $T(f)$ is clear; its continuity results from Lebesgue dominated convergence theorem.

Making use of some basic facts from integration theory, it is possible to conclude that

[3] The notion of continuity hereby employed is that one based on sequences, as in normed spaces, and of course based on the convergence notion introduced on $\mathcal{D}(\Omega)$.

$$f, g \in L^1_{\text{loc}}(\Omega), T(f) = T(g) \Longrightarrow f = gae,$$

cf. [35, pp. 288]. The notation $T_f := T(f)$ may as well be employed.

Example 4.3 is the inspiration to – informally – express that

every function is a distribution.

Observe that the converse fails to hold: for example,

$$_{\text{loc}} f \in L^1_{\text{loc}}(\Omega)_{\text{loc}} T(f) = \delta_{\bar{x}}.$$

The distribution $\delta_{\bar{x}}$ is called Dirac delta "function." Due to the success drawn from the way some physicists and engineers dealt with "functions" like this one, cf. Sect. 3.8, some mathematicians[4] were pushed to find the right theoretical framework to understand them.

Despite not every distribution being defined through a given function according to (4.16), it is possible to introduce some concepts and operators on $\mathcal{D}'(\Omega)$ that in principle make only sense for functions. This turns out to be a framework which is quite similar to that one described in Sect. 2.14, to study the "generalized functions" from $L^1(I\!R)$. The distributions are sometimes also called **ideal functions** or **generalized functions**, cf. [19, 33].

Given $\psi \in C^\infty(\Omega)$ and $T \in \mathcal{D}'(\Omega)$, the product of a function by a distribution is defined by $(\psi T) \in \mathcal{D}'(\Omega)$:

$$< \psi T, \phi >:=< T, \psi\phi >, \forall \phi \in \mathcal{D}(\Omega). \tag{4.17}$$

Observe that when $T = T(f)$, we obtain $\psi T = T(f\psi)$, that is, when distributions given by functions are considered, the product defined by (4.17) comes to be the usual function product.

By the same token, let $a \in I\!R^n$ and $f \in L^1_{\text{loc}}(I\!R^n)$. Denoting by $\tau^a f$ the shift of f, that is,

$$(\tau^a f)(x) := f(x - a),$$

we get, for $\phi \in \mathcal{D}(I\!R^n)$,

$$\begin{aligned} < T(\tau^a f), \phi > &= \textstyle\int_{I\!R^n} f(x-a)\phi(x)dX = \\ &\textstyle\int_{I\!R^n} f(x)\phi(x+a)dX = < T(f), \tau^{-a}\phi >, \end{aligned}$$

which motivates to pose the definition for **distribution shift**, through:

$$< \tau^a T, \phi >:=< T, \tau^{-a}\phi >, \phi \in \mathcal{D}(I\!R^n).$$

---

[4] Among them Sobolev, Friedrichs, Schwartz, and Gel'fand.

Analogously, the formula of integration by parts (4.3) gives, for $f \in C^k(\Omega)$,

$$\begin{aligned} < T(D^\alpha f), \phi > &= \int_\Omega \phi(D^\alpha f)dX \\ &= (-1)^{|\alpha|} \int_\Omega f(D^\alpha \phi)dX = (-1)^{|\alpha|} < T(f), D^\alpha \phi >, \end{aligned}$$

for any $\phi \in \mathcal{D}(\Omega)$ and $|\alpha| \leq k$. Define, for α an arbitrary multi-index,

$$< D^\alpha T, \phi > := (-1)^{|\alpha|} < T, D^\alpha \phi > \quad \begin{matrix} \forall \phi \in \mathcal{D}(\Omega) \\ \forall T \in \mathcal{D}'(\Omega) \end{matrix} . \tag{4.18}$$

Therefore, a distribution exhibits derivatives of all orders.

Given $f \in L^1_{\text{loc}}(\Omega)$, it is said that $D^\alpha T(f)$ is the **derivative of f in the sense of distributions** . If f has a derivative in the classical sense, or on the strong sense, they both coincide with its derivative in the sense of distributions.

Example 4.4 Suppose that $T \in \mathcal{D}'(I\!R)$ satisfies $DT = 0$. Then T is a constant, *i.e.*, $T = T_s$, for some constant function s.

Indeed, by (4.18), for any $\theta \in \mathcal{D}(I\!R)$, we have that

$$0 =< DT, \theta >= - < T, \theta' > .$$

But Exercise 3.2(ii) states that, given $\phi \in \mathcal{D}(I\!R)$, the identity

$$\phi = \theta' + \alpha\phi_0$$

holds, being $\theta \in \mathcal{D}(I\!R)$ and $\alpha \in I\!R$ that depend on ϕ, while $\phi_0 \in \mathcal{D}(I\!R)$ is a fixed function. Let us be explicit, $\alpha := (\int \phi)/(\int \phi_0)$. Thus,

$$< T, \phi >=< T, \theta' + \alpha\phi_0 >= \alpha < T, \phi_0 >= \int_{I\!R} s(x)\phi(x)dx,$$

for

$$s(x) :=< T, \phi_0 > / \int_{I\!R} \phi_0(t)dt = \text{[some constant]} .$$

Exercise 4.10 (a) Calculate the derivative of order k of the distribution associated to the **Heaviside function**

$$H(x) := \begin{bmatrix} 0 & x < 0 \\ 1 & x > 0 \end{bmatrix} .$$

(b) If $\iota := (1, 2, \ldots, n)$, calculate $D^\iota H_n$, for

$$H_n(x) := H(x_1)H(x_2)\ldots H(x_n).$$

Exercise 4.11 Verify that $f(x) := \ln|x| \in L^1_{\text{loc}}(I\!R)$ and calculate the first derivative of $T(f)$.

We now define a notion of convergence in $\mathcal{D}'(\Omega)$, the so-called **pointwise convergence of distributions**. It is said that the sequence $T_J \in \mathcal{D}'(\Omega)$ converges to $T \in \mathcal{D}'(\Omega)$ if, for each $\phi \in \mathcal{D}(\Omega)$, we have

$$\lim_J < T_J, \phi > = < T, \phi > .$$

It is worth comparing the results about termwise differentiation of sequences of functions with

Theorem 4.2 *For any sequence $T_J \in \mathcal{D}'(\Omega)$, with $T_J \stackrel{J}{\to} T \in \mathcal{D}'(\Omega)$ and any choice of multi-indices α, the convergence*

$$D^\alpha T_J \stackrel{J}{\to} D^\alpha T,$$

holds, i.e., any convergent sequence of distributions may (validly) go through termwise differentiation.

Indeed, let $\phi \in \mathcal{D}(\Omega)$ and α be a multi-index. We then have

$$< D^\alpha T_J, \phi > =$$

$$(-1)^\alpha < T_J, D^\alpha \phi > \stackrel{J}{\to} < T, D^\alpha \phi > =$$

$$= < D^\alpha T, \phi >,$$

and this, by definition, means that $T_J \to T$.

Banach-Steinhaus theorem, with a formulation slightly more general than the one described on Chap. 5, implies

Theorem 4.3 *Suppose that $T_J \in \mathcal{D}'(\Omega)$ is such that, for each $\phi \in \mathcal{D}(\Omega)$, there exists the* $\lim_{J\to\infty} < T_J, \phi >$. *Then*

$$< T, \phi > := \lim_{J\to\infty} < T_J, \phi >$$

defines a distribution in $\mathcal{D}'(\Omega)$.

4.5 Vector Functions and Distributions

To study the evolution equations, a generalization of the concepts just introduced is needed.

Consider B as a Banach space with norm $\|\cdot\|_B$ and, for fixed $T \in (0, \infty)$ and $1 \leq p < \infty$, define the space $L^p(0, T; B)$ as the completion of

$$C^0(0, T; B) := \{f : [0, T] \to B;\ \text{for continuous} f\},$$

equipped with the norm

$$|f|_{L^p(0,T;B)} := \left[\int_0^T \|f(t)\|_B^p dt\right]^{1/p}, 1 \leq p < \infty.$$

(It ought to be considered that, when $T = \infty$, the completion must be taken not for $C^0(0, T; B)$ but, instead, for the space $C_0^0(0, T; B)$.)

Essentially, all results that hold for the spaces of real functions, *i.e.*, when $B = I\!R$, remain valid within the new context. In particular, we have Minkowski inequality as well as the dual representations:

$$\left[\int_0^T \|f+g\|_B^p dt\right]^{1/p} \leq \left[\int_0^T \|f\|_B^p dt\right]^{1/p} + \left[\int_0^T \|g\|_B^p dt\right]^{1/p}, \tag{4.19}$$

for arbitrary $f, g \in L^p(0, T; B)$, with $1 \leq p < \infty$;

$$[L^p(0, T; B)]^* = [L^q(0, T; B^*)], q := (1 - 1/p)^{-1}, 1 \leq p < \infty. \tag{4.20}$$

Expression (4.20) means that, given $\ell \in [L^p(0, T; B)]^*$, there exists a unique $G = G_\ell \in L^q(0, T; B^*)$ for which

$$\ell(f) = \int_0^T G(t) \cdot f(t) dt, \forall f \in L^p(0, T; B), \tag{4.21}$$

where $G(t) \cdot f(t)$ stands for the duality, quite often denoted by $< G(t)|f(t) >$.

Note that Hölder inequality implies the existence of the integral in (4.21): from

$$|G(t) \cdot f(t)| \leq \|G(t)\|_{B^*} \|f(t)\|_B$$

it follows that

$$\left|\int_0^T G(t) \cdot f(t) dt\right| \leq |G|_{L^q(0,T;B^*)} |f|_{L^p(0,T;B)}. \tag{4.22}$$

Expression (4.22) may be thought of as Hölder inequality, throughout the current framework.

In a similar fashion, if B is a Hilbert space with inner product $(\cdot|\cdot)_B$, then $L^2(0, T; B)$ is also a Hilbert space when equipped with the inner product

$$< f(t)|g(t) >:= \int_0^T (f(t)|g(t))_B \, dt,$$

being Schwarz inequality then written as

$$< f|g > \leq \|f\|_{L^2(0,T;B)} \|g\|_{L^2(0,T;B)}.$$

The latecomer $L^\infty(0, T; B)$ is introduced as the space of all functions f that map $(0, T)$ on B, which are measurable and that hold the property:

$$\sup_t \text{ess} \|f(t)\|_B < \infty.$$

(A function $f : [0, T] \to B$ is said to be **measurable** if it happens to be the *ae* limit of a sequence of functions on $L^1(0, T; B)$.)

The **vector-valued distributions** (on B) are the linear continuous operators from $\mathcal{D}(0, T)$ to B. The space of those distributions is denoted by $\mathcal{D}'(0, T; B)$.

The functions f on $L^p(0, T; B)$ are associated in a natural fashion to the distributions from $\mathcal{D}'(0, T; B)$ through the correspondence

$$\begin{array}{llll} \tau : L^p(0, T; B) \to \mathcal{D}'(0, T; B) & & & \\ \quad f \quad \to \quad T(f) & : \mathcal{D}(0, T) \to B & & \\ & \quad \phi \quad \to < T(f), \phi >:= & & \\ & \quad \left[\int_0^T \phi(t) f(t) dt\right] & & \end{array} \tag{4.23}$$

where the meaning of the integral is given by the same procedure employed for real valued functions: the **continuous extension** of the Riemann integral operator defined on $C_0(0, T; B)$.

The notions like those of derivative and convergence on $\mathcal{D}'(0, T; B)$ are as well introduced following the same track as those on $\mathcal{D}'(0, T)$. We shall spare the reader from them, but not from the well recognized as important

Theorem 4.4 *Being defined, for any* $f \in L^p(0, T; B), 1 \leq p < \infty$, *the distribution* $T(f) \in \mathcal{D}'(0, T; B)$ *by (4.23), for it always hold:*

(i) If $f_1, f_2 \in L^p(0, T; B)$ *satisfy*

$$< T(f_1), \phi > = < T(f_2), \phi >, \forall \phi \in \mathcal{D}(0, T),$$

then

$$f_1 = f_2;$$

(ii) If $f_j, f \in L^p(0, T; B)$, and, for any $G \in L^q(0, T; B*)$, we have that[5]

$$\int_0^T G(t) \cdot [f_j(t) - f(t)]dt \to 0,$$

then

$$< T(f_j), \phi > \to < T(f), \phi >, \forall \phi \in \mathcal{D}(0, T), \tag{4.24}$$

or, in an alternate saying,

$$T(f_j) \to T(f) \textit{ in the sense of } \mathcal{D}'(0, T; B).$$

A sequence f_j for which (4.24) holds is said to converge to f **in the sense of the distributions.**

4.6 The Trace Theorem

The elements in $H^k(\Omega)$ belong to $L^2(\Omega)$. Due to such pedigree, they turn out to be associated to distributions, and those have derivatives of every order. But it then occurs that the derivatives of order $\leq k$ of these distributions are given precisely by the derivatives in the strong (or weak) sense introduced in Sects. 4.2 and 4.3. In other words, it is among our tools the

Theorem 4.5 *The space $H^k(\Omega)$ may be defined as the set of functions from $L^2(\Omega)$ whose derivatives in the sense of distributions, of order $\leq k$, belong to $L^2(\Omega)$.*

Let Ω be a bounded domain. Our purpose now is to characterize $H_0^k(\Omega)$.

Observe that, based on the generic characterization for $H^k(\Omega)$ exposed above, by which it is meant a set of functions from $L^2(\Omega)$ with additional properties, it makes no sense to look for the values of a given function $f \in H^1(\Omega)$ on $\partial\Omega$. In fact, besides being f defined *ae*, this is added to the fact that $\partial\Omega$ is a null measure set. And, if worse gets to worst, nothing was required from f with regard to $\partial\Omega$, while defining $H^1(\Omega)$ – recall, as the $C_*^1(\Omega)$ completion.

[5] It is said then that the sequence f_j converges weakly to the function f. This concept will be introduced in a more general – as well as more precise – fashion in Sect. 5.4.

Let us make a tour back on Sect. 2.10. We have reached therein, by putting to work the inequality (2.26), that for every $f \in C^1_*(0,1)$,

$$|f(t_1) - f(t_2)| \leq \sqrt{|t_1 - t_2|}\|f\|_{1,2} \tag{4.25}$$

holds, and since, due to (2.14),

$$\|f\|_{0,\infty} \leq \sqrt{2}\|f\|_{1,2} \tag{4.26}$$

is true, we may deduce that the functions from $H^1(0,1)$ may be thought as being uniformly continuous. The real meaning of this assertion is that, as long as the elements from $H^1(0,1)$ are equivalence classes of functions, given any of these classes, it contains a uniformly continuous function. In alternate terms, given a function from $H^1(0,1)$, it is possible to modify its values on a null measure set in such a way as to make it a uniformly continuous function.

For $f \in C^1_*(0,1)$, define the **trace** of f, denoted by γf, as the operator which generates (ou retrieves) the values of f on the boundary of $[0,1]$. It may be deduced that it is possible to define γ as a bounded operator on $H^1(0,1)$:

(i) By (4.25), every f in $C^1_*(0,1)$ is uniformly continuous on $(0,1)$, thus may be extended to $[0,1]$.
(ii) The operator $f \mapsto \gamma f$ is linear and, by (4.26), bounded, therefore uniformly continuous on $C^1_*(0,1)$; moreover it may be (uniquely) extended to the completion of $C^1_*(0,1)$, which is $H^1(0,1)$.

Finally, observe that on $H^1_0(0,1)$ it must necessarily be the null operator. Indeed, γ is null on$C^1_0(0,1)$ and – due to the continuity – on its closure, which turns out to be exactly $H^1_0(0,1)$.

This example was presented as to motivate the trace theorem, described in the sequel. Before that we need a definition:

A function f is said to belong to $C^k(\overline{\Omega})$ if it admits an extension $\tilde{f} \in C^k(\tilde{\Omega})$, where the open set $\tilde{\Omega} \supset \overline{\Omega}$.

Trace Theorem *Let Ω be an open bounded set, whose boundary $\partial\Omega$ is a surface of class C^2. Then there exists a unique linear operator*

$$\gamma : H^1(\Omega) \to L^2(\partial\Omega),$$

called ***trace operator****, for which whenever $f \in C^1(\overline{\Omega})$, we have that $\gamma f = f|_{\partial\Omega}$.*

Furthermore:

(i) $\|\gamma f\|_{L^2(\partial\Omega)} \leq C\|f\|_{H^1(\Omega)}$
(ii) $ker(\gamma) = H^1_0(\Omega)$

A remark is worth: it can be verified that the operator γ fails to be onto $L^2(\Omega)$. In fact, $\gamma\left(H^1(\Omega)\right) = H^{1/2}(\partial\Omega)$, where $H^{1/2}(\partial\Omega) \subset L^2(\partial\Omega)$ is one of the so-called Sobolev spaces of fractional order (see the following section).

References for this theorem are [5, 50, 51], the last one in Portuguese.

A formulation of the trace theorem for vector valued functions is required for dealing with evolution partial differential equations.

4.7 Sobolev Spaces of Real Order

Sobolev spaces $H^s(\Omega)$, for s **integer**, have been defined and some of their properties discussed on previous sections. When the trace theorem was formulated, we had to mention the space $H^{1/2}(\Omega)$, which gives an idea of how we may need to generalize these spaces for an arbitrary **real** s. That is the aim of the discussion on this section.

4.7.1 **δ-Function** *Representations*

Consider for $0 < a \leq 1$ the function ψ_a defined by

$$\psi_a(x) = \begin{bmatrix} \cosh x/\sinh a & 0 < x < a \\ 0 & a < x < 1 \end{bmatrix}.$$

It is seen that the functional associated to ψ_a by

$$\ell_{\psi_a}(f) := <\psi_a|f>_{H^1} = \int_0^1 \psi_a f + \int_0^1 \psi_a' f',\ f \in H^1(0,1) \tag{4.27}$$

turns out to be the **Dirac function** $\delta(x-a)$ introduced[6] on Sect. 4.4. Better said, ψ_a is the representation in H^1 of the functional δ_a, explained with Riesz representation theorem.

It became a piece of the mathematical folklore to refer to the *ghost function* δ as a mathematical concept (intrinsically) defined through the property

$$\int \delta(x-a) f(x) dx = f(a), \forall f, \tag{4.28}$$

despite not a smart enough **function** exists as to fulfill simultaneously (4.28), for any f, as long as it is being considered Riemann or even Lebesgue integral. (This claim is true because (4.28) implies that $\delta(x) = 0ae$, from which it follows the

---

[6] Already mentioned on Example 2.12b, Sect. 2.5.

vanishing of the integrals of $\delta \times f$, for any f. Employing then easy to operate dense subsets is the way to make a clear reasoning.)

The representation for δ_a described in (4.27) gets to be coherent with (4.28), though, as long as we treat the integral in that expression as a notation for the inner product in H^1, but not in L^2.

Observe now that, besides this identification

$$\psi_a(x) \longleftrightarrow \delta(x-a) \text{ on } H^1(0,1),$$

as ψ (forget for a while the index a) belongs to $L^2(0,1)$, it is also associated to the linear functional given by

$$\begin{aligned} \mathcal{L}_\psi : L^2(0,1) &\to I\!R \\ f &\to \mathcal{L}_\psi(f) := \int_0^1 \psi(x) f(x)dx =< \psi | f >_{L^2} \end{aligned} . \tag{4.29}$$

Now, take into account that the expression (4.29) even defines another linear continuous functional on $H^1(0,1) \subset L^2(0,1)$, which will be denoted by $\lambda = \lambda_\psi$. And besides, for such functional the following inequality holds:

$$|\lambda(f)| \leq \|\psi\|_{L^2}\|f\|_{L^2} \leq \|\psi\|_{L^2}\|f\|_{H^1},\ f \in H^1(0,1). \tag{4.30}$$

As a consequence of those inequalities, it can be deduced that

$$\|\lambda_\psi\| := \sup_{\|f\|_{H^1}=1} |\lambda(f)| \leq \|\psi\|_{L^2} = \|\mathcal{L}_\psi\|. \tag{4.31}$$

In all the above steps, ψ may be replaced by an arbitrary function on $L^2(0,1)$ – that's the way we have reached an identification, denoted by $\mathcal{I}$, of $L^2(0,1)$ with a part of the dual of $H^1(0,1)$, which may as well get the formulation herein described.

As long as $H^1 \subset L^2$ and being continuous[7] the identity $I\!I : H^1 \to L^2$, it is deduced that $(L^2)^* \subset (H^1)^*$. The identification $\mathcal{I}$ follows from composing the mapping $\imath$, given by Riesz representation theorem in L^2, with the identity $I\!I$ – now from $(L^2)^*$ to $(H^1)^*$:

$$L^2 \xrightarrow{\ \imath\ } (L^2)^* \xrightarrow{\ I\!I\ } (H^1)^*, \qquad L^2 \xrightarrow{\ \mathcal{I}\ } (H^1)^*$$

[7] This property is in general expressed by saying that "H^1 **is continuously immersed in** L^2".

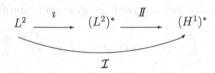

Let us assign the notation $\mathcal{L}^2$ to represent the vector space associated to the elements of L^2, but putting aside the norm $\| \cdot \|_2$. The just described procedure allows to introduce in $\mathcal{L}^2$ **another** norm, denoted by $\| \cdot \|_{-1}^{\tilde{}}$, namely:

Given $\psi \in \mathcal{L}^2$, its norm gets **defined** as the norm of the functional $\lambda_\psi \in (H^1)^*$, cf. (4.31).

From (4.31) it may be deduced that

$$\|\psi\|_{-1}^{\tilde{}} \leq \|\psi\|_{L^2}, \, \psi \in \mathcal{L}^2(0,1), \tag{4.32}$$

an inequality which naturally leads to question whether these two norms on $\mathcal{L}^2$ are equivalent, which means:

Is it possible to claim the existence of $\alpha > 1$ such that

$$\|\psi\|_{L^2} \leq \alpha \|\psi\|_{-1}^{\tilde{}}, \tag{4.33}$$

for every $\psi \in \mathcal{L}^2(0,1)$?

We call for help from the *delta function* to reach the answer.

Claim There exist functions $\psi_n \in \mathcal{L}^2(0,1)$ such that

$$\limsup \|\psi_n\|_{-1}^{\tilde{}} \leq 1 \text{ but } \|\psi_n\|_{L^2} \to \infty \text{ if } n \to \infty. \tag{4.34}$$

Proof Take

$$\psi_n(x) := \begin{bmatrix} n & 0 < x < 1/n \\ 0 & x \geq 1/n \end{bmatrix} \quad n > 0, \tag{4.35}$$

so that

$$\|\psi_n\|_{L^2} = \sqrt{n}. \tag{4.36}$$

On the other hand,

$$\lambda_{\psi_n}(f) := \int_0^1 \psi_n f = n \int_0^{1/n} f \to f(0), \, f \in H^1(0,1),$$

since we have that $H^1(0,1) \subset C^0([0,1])$. Consequently, the sequence ψ_n is bounded with respect to the norm $\|\psi\|_{-1}^{\tilde{}}$, because we can observe that

$$\limsup \|\psi_n\|_{-1}^{\tilde{}} \leq \sup\{|f(0)|; \, \|f\|_{H^1} = 1\} \leq 1, \tag{4.37}$$

and from the use of (2.14), it is seen that

$$\|f\|_\infty \leq \sqrt{2}\|f\|_{H^1} \text{ in } H^1(0,1).$$

Expressions (4.36) and (4.37), that conduct to prove (4.34), contradict to be possible (4.33) to hold.

An additional information which can be deduced from this result is that the set $\mathcal{L}^2$, if equipped with the norm $\|\cdot\|_{-1}^{\tilde{}}$, fails to be a complete space. This turns out to be a consequence of Theorem 2.2 from Sect. 2.13, whose statement we recall.

Given a Banach space B and a norm which turns out to be comparable to its original one, these norms are equivalent if and only if B remains complete when equipped with the latter norm.

A more constructive proof follows. The sequence $\{\psi_n\}$ from (4.35) is a Cauchy sequence with respect to the norm $\|\cdot\|_{-1}^{\tilde{}}$, but it does not converge in $\mathcal{L}^2$ for this same norm. The relations

$$\begin{aligned}
\|\psi_n - \psi_m\|_{-1}^{\tilde{}} &= \sup_{\|f\|_{H^1}=1} \int f(\psi_n - \psi_m) \\
&= \sup_{\|f\|_{H^1}=1} \left\{ n\int_0^{1/n} f - m\int_0^{1/m} f \right\} = \sup_{\|f\|_{H^1}=1} \{f(x_n) - f(x_m)\} \\
&\leq \sup_{\|f\|_{H^1}=1} |x_n - x_m|^{1/2}\|f'\|_{L^2} \leq |x_n - x_m|^{1/2} \leq [\max\{1/n, 1/m\}]^{1/2},
\end{aligned}$$

– where the first inequality comes from (2.26) – lead to the reasoning that follows.

Assume a certain $g \in \mathcal{L}^2(0,1)$ satisfy $\|\psi_n - g\|_{-1}^{\tilde{}} \to 0$. Then it follows that, for every $f \in H^1(0,1)$, it would be true that

$$\begin{aligned}
0 = \lim < \psi_n - g|f >_{L^2} &= \lim < \psi_n|f >_{L^2} - < g|f >_{L^2} \\
&= \delta(f) - < g|f >_{L^2} .
\end{aligned} \tag{4.38}$$

As a consequence, g ought to satisfy

$$\int g(x)f(x) = f(0), \forall f \in H^1(0,1). \tag{4.39}$$

But this can not hold, as we already know, a proof suggested after (4.28).

Recall that the set $\mathcal{L}^2(0,1)$ has been obtained as a **completion**. At the present stage, with this new norm, known as Lax **negative norm** , it is possible to get again its completion:[8] we reach then a space denoted[9] by $\tilde{H}^{-1}(0,1)$. The following relations are then true:

[8] Compare this step with the result on Exercise 2.25.

[9] We must emphasize that this one is not a standard notation, as opposed to $H^{-1}(0,1)$, cf. Sect. 3.

$$\tilde{H}^{-1}(0,1) \supset H^0(0,1) = L^2(0,1) \supset H^s(0,1), s \geq 1, \tag{4.40}$$

as well as the **generalized Schwarz inequality**

$$\int fg \leq \|f\|_{-1}^{\sim} \|g\|_{H^1}, \ f \in \tilde{H}^{-1}(0,1), g \in H^1(0,1), \tag{4.41}$$

which is deduced from (4.31), as follows:

$$< f|g/\|g\|_{H^1} >= \lambda_f(g/\|g\|_{H^1}) \leq \|\lambda_f\| =: \|f\|_{-1}^{\sim}.$$

By construction, the space $\tilde{H}^{-1}(0,1) \subset [H^1(0,1)]^*$. It is then natural to question:

> How far is $\tilde{H}^{-1}$ from being equal to the whole of $(H^1)^*$?

Otherwise:

> Did we happen to have constructed a (new) representation for $(H^1)^*$? If not yet, what is missing, then?

Still in an alternate way:

> Which are the functionals on H^1 which can not be approximated by functions from L^2, by means of the identification ℓ_ψ described in (4.27)?

Consider once more the sequence $\psi_n \in \mathcal{L}^2$. The relation (4.38), which points up to the impossibility of its convergence in $\mathcal{L}^2$, also shows that, when looked at as a sequence of linear functionals, it is **pointwise convergent** to the *delta function*.

Examples similar to this one are quite spread through the (not too mathematically rigorous) literature, being presented as *sequences that converge to the delta function*. Their basis is the fact that the functions ψ_n fulfill

$$\left.\begin{array}{ll} \lim_n \psi_n(x) = 0 & x \neq 0 \\ \lim_n \psi_n(x) = \infty & x = 0 \\ \lim_n \int \psi_n(x) f(x) dx = f(0) \ \forall f & \end{array}\right],$$

while it is assumed that the *delta function* satisfies

$$\left.\begin{array}{ll} \delta(x) = 0 & x \neq 0 \\ \delta(x) = \infty & x = 0 \\ \int \delta(x) f(x) dx = f(0) \ \forall f & \end{array}\right].$$

In the sequel, it is verified the possibility to reach more deep convergence results. As long as we get aware of the pointwise convergence of the linear functionals associated to the sequence ψ_n, we are able to claim that such a limit is also a linear functional which lives in the same space.[10] Saying it with another words, if we complete $\mathcal{L}^2$, equipped with the norm $\|\cdot\|_{-1}^{\sim}$, we shall reach the functionals δ that, as is already known, do not belong to the L^2 space. This has brought as in shape to pose the conjecture which rests on

$$\textbf{Theorem 1} \tilde{H}^{-1}(0,1) = [H^1(0,1)]^*. \tag{4.42}$$

Proof The claim in (4.42) amounts to show that $\mathcal{I}(L^2)$ is dense in $(H^1)^*$. And this means: given a functional T on $(H^1)^*$ –, *i.e.*, $T \in (H^1)^{**}$ – which vanishes on $\mathcal{I}(L^2)$, we ought to have $T = 0$.

Now, knowing that H^1 is reflexive, it is deduced that such an operator T is the image of some $\phi \in H^1$ by the canonical isomorphism $T = T_\phi$ where:

$$T(\ell) = \ell(\phi), \forall \ell \in (H^1)^*. \tag{4.43}$$

It must be pointed that, whenever we take $\ell \in \mathcal{I}(L^2)$, we get $\ell = \ell_\psi$ for a certain $\psi \in L^2$, and therefore (4.43) linked to the hypothesis about T leads to

$$0 = T(\ell) =< \psi|\phi >_{L^2}, \forall \psi \in L^2. \tag{4.44}$$

In short, $\phi \in H^1 \subset L^2$ is orthogonal to L^2, therefore null, which closes the proof.

It is worth having this result rewritten as

Theorem 1' *Any functional from* $H^1(0,1)^*$ *may be approximated* ***with respect to the operator norm*** *by functionals of type*

$$\left.\begin{array}{ll} \ell_\psi : H^1(0,1) \to I\!R \\ \quad f \quad \to \ell_\psi(f) := \int f\psi \end{array}\right] \psi \in L^2(0,1).$$

When this result gets joined to Riesz theorem, we reach the two following representations for the dual of $H^1(0,1)$:

For every functional $\ell \in H^1(0,1)^*$ and each $\epsilon > 0$, there exist functions $\psi_\epsilon \in L^2(0,1)$ and $\phi \in H^1(0,1)$ – the latter uniquely determined – for which the following identity and inequality hold

$$\left.\begin{array}{c} \ell(f) = < f|\phi >_{H^1(0,1)} \\ \left|\ell(f) - < f|\psi_\epsilon >_{L^2(0,1)}\right| < \epsilon \|f\|_{H^1} \end{array}\right] f \in H^1(0,1). \tag{4.45}$$

It is quite important to keep in mind that $\tilde{H}^{-1}$ and H^1 are two distinct spaces, but they are isometrically identified to the same space $(H^1)^*$. Due to these **two**

[10] A consequence of the Banach-Steinhaus theorem, Sect. 5.5, Exercise 5.11.

identifications, $(H^1)^*$ inherits the inner product from H^1 (Riesz representation theorem) and exports it – via (4.29) – to the space $\tilde{H}^{-1}$.

This subsection is shut with the dual variational formulas

$$\|\psi\|\tilde{_{-1}} = \sup\{<\psi, f>; \|f\|_{H^1} \leq 1\}, \psi \in \tilde{H}^{-1}(0,1) \tag{4.46}$$

$$\|f\|_{H^1} = \sup\{<\psi, f>; \|\psi\|\tilde{_{-1}} \leq 1\}, f \in H^1(0,1) \tag{4.47}$$

where

- $<\psi, f>$ denotes the action of the functional $\psi \in \tilde{H}^{-1}(0,1) = (H^1)^*$ on f – usually quoted as **duality**
- The identity (4.46) repeats (4.31)
- And, finally, identity (4.47) is a consequence of Hahn-Banach theorem – to be discussed on the following chapter, check further Exercises 3.3 and 5.2 and Sect. 5.3.

Exercise 4.12 $\left[\tilde{H}^{-1}(0,1)\right]^* = H^1(0,1)$.

Hint For $f \in H^1$,

$$\begin{aligned} \ell_f : \tilde{H}^{-1} &\to I\!R \\ \psi &\to \ell_f(\psi) := <\psi, f> \end{aligned}$$

defines an isometry. In order to conclude it is onto $(\tilde{H}^{-1})^*$, the same reasoning employed for Theorem 1 may be recalled.

4.7.2 The Dual Space $H^{-1}(0,1)$

Instead of working on $H^1(0,1)$ as in the previous section, choose now $H_0^1(0,1)$: all steps may be repeated, the set $\mathcal{L}^2(0,1)$ being equipped at this turn with the norm

$$\|\psi\|_{-1} := \sup\{<\psi|f>_{L^2}; f \in H_0^1(0,1), \|f\|_{H^1} \leq 1\}. \tag{4.48}$$

Its completion is then denoted by $H^{-1}(0,1)$. The relations deduced for $\|\cdot\|\tilde{_{-1}}$ as well as $\tilde{H}^{-1}$ remain valid for $\|\cdot\|_{-1}$ and H^{-1} whenever exchange is done of H^1 by H_0^1. It is worth to emphasize (4.31)–(4.32)–(4.40)–(4.41)–(4.42)–(4.46)–(4.47), as well as Theorem 2.1 and Exercise 2.1.

After all that work called for constructing $\tilde{H}^{-1}$, the expressions in (4.45) may be rated as unsatisfactory, as long as the reached identification is a bit loose: while the former adds nothing to what Riesz representation theorem has already put on our hands, the latter just rewrites information previously available about the completion technique. We thus would struggle to exhibit a more concrete characterization for its elements. We fail to reach that for $\tilde{H}^{-1}$, nevertheless for H^{-1}, we were able to arrive at something more tangible.

In fact, let $\psi \in H^{-1}(0,1) = [H_0^1(0,1)]^*$ be an arbitrary functional. Riesz representation theorem assures the existence of a unique $g \in H_0^1(0,1)$ for which

$$< \psi, f >=< g|f >_{H^1}, \, f \in H_0^1(0,1),$$

which means

$$< \psi, f >= \int_0^1 gf + \int_0^1 g'f', \, f \in H_0^1(0,1).$$

Suppose that f allows integration by parts

$$< \psi, f >= \int_0^1 gf - \int_0^1 gf'', \tag{4.49}$$

an expression which motivates to take[11] $f \in C_0^\infty(0,1) = \mathcal{D}(0,1)$ and have it rewritten as

$$< \psi, f >=< T(g), f > - < D^2T(g), f > . \tag{4.50}$$

Denote by $T(g)$ the distribution from $\mathcal{D}'(0,1)$ associated to the function g and by $D^2T(g)$ its second derivative – cf. (4.16) and (4.18). (Beware, the brackets stand, in the right-hand side, for duality $< \mathcal{D}'(0,1), \mathcal{D}(0,1) >$, and on the left-hand side, analogously for $< H^{-1}, H_0^1 >$.) The element $\psi \in H^{-1}$ may thus be **identified** to the distribution $S := [T(g) - D^2T(g)] \in \mathcal{D}'(0,1)$, since it is the **unique** extension of S to H_0^1.

The above remarks thus let us to think on H^{-1} as a space of distributions obtained from H_0^1 by the operator

$$I - \partial^2/\partial x^2 : H_0^1 \to H^{-1}, \tag{4.51}$$

being the derivative $\partial^2/\partial x^2$ understood within the framework just described: the extension to H_0^1 of the derivative of a distribution. The operator in (4.51) is a representation of the canonical isomorphism between H_0^1 and its dual H^{-1}, which from now on we start to refer as a space of distributions. A conclusion reached at first by adapting (4.45) to H_0^1 and H^{-1} may be deleted.

[11] The reader should spare the notation looseness in this equality, which must be thought relating strictly the elements of the mentioned sets, no mention being made upon their structures.

Observe that for H^1 we can not reach a representation as simple as this one. Indeed, besides being compressed by requirements from (4.49) – boundary terms must be dealt with – it shows up to be needed by the above reasoning that $\mathcal{D}(0, 1)$ is dense in $H_0^1(0, 1)$, so that uniqueness for the extension for S rests assured.

In order to be led to still another representation for H^{-1}, the following generalization of (4.29) is called:

Given $v_1, v_2 \in L^2$, taking $v := (v_1, -v_2) \in L^2 \times L^2$ and

$$\begin{aligned} L_v : H_0^1 &\to I\!R \\ f &\to < L_v, f > := \int f v_1 - \int f' v_2 \end{aligned},$$

we identify L_v to a distribution from H^{-1}, namely,

$$L_v \longleftrightarrow v_1 + v_2', \, v_1, v_2 \in L^2.$$

From this reasoning we can conclude that every element of $L^2 \times L^2$ generates a distribution from H^{-1}, throughout a mapping that clearly fails to be 1-1, since $L_v = 0$ when taking $v := (-u', u)$, for any $u \in L^2$ for which $u' \in L^2$ also holds. But this mapping happens to be onto, as taught by the following

Theorem 4.6 *Given arbitrary $\psi \in H^{-1}$, it is possible to determine a pair $v_1, v_2 \in L^2$ such that, on the sense of the distributions, we can count on the identity*

$$\psi = v_1 + v_2'. \tag{4.52}$$

Besides, it is always possible to choose v_1, v_2 such that

$$\|\psi\|_{-1} = \{\|v_1\|_{L^2}^2 + \|v_2\|_{L^2}^2\}^{1/2}.$$

Proof Any $\psi \in H^{-1}$ is naturally identified with a functional λ, defined on the graph $G \subset L^2 \times L^2$ of the operator

$$d/dx : H_0^1 \to L^2,$$

taking into account that this subspace is isometrically isomorphous to H_0^1. The functional λ may be extended by, say, $\tilde{\lambda}$ to the whole space $L^2 \times L^2$, in such a way that the norm of λ is preserved. Besides, for it the representation

$$\tilde{\lambda}(w) = \int v_1 w_1 + \int v_2 w_2, \forall (w_1, w_2) \in L^2 \times L^2$$

holds, being $v = (v_1, v_2) \in L^2 \times L^2$ uniquely determined.

In particular, for arbitrary $\omega \in H_0^1$, we have $w := (\omega, \omega') \in G$ and

$$< \psi, \omega >= \lambda(w) = \int v_1\omega + \int v_2\omega', \omega \in H_0^1. \tag{4.53}$$

Observe that the expression

$$\|\psi\|_{-1} = \|\lambda\| = \|\tilde{\lambda}\| = \|v\|_{L^2 \times L^2}$$

holds and that the reasoning so far presented may be applied as well to H^1, *i.e.*, the representation (4.53) holds for both H^1 and H_0^1. But it should be emphasized that only for the latter it is truly correct to pass from (4.53) to (4.52) the same reasoning line built before.

Observation Given $\psi \in H^{-1}$, we have remarked the validity of (4.53). By building the distribution $T(v_1) + DT(v_2)$, it allows a continuous extension (which may fail to be unique) to H^1, which lets to express the fact that $\tilde{H}^1$ is the space of extensions of distributions of type (4.52). In brief, it is not possible to reach the expression (4.52) for all functionals from $\tilde{H}^1$ due to the boundary terms. In spite of that, we can reach the following representation – cf.[5], p. 60:

$$< \psi, w >= \int w_1v_1 + \int w_1'v_1 + \alpha_0 w(0) + \alpha_1 w(1),$$

with $v_1, v_2 \in L^2$ and $\alpha_0, \alpha_1 \in I\!R$.

4.7.3 The Spaces H^{-p}, *p* Integer

Let us put together now the three ways to look at the space $H^{-1}(0, 1)$, but we will make the choice of already presenting them for a natural generalization, namely, for $H^{-p}(0, 1)$, $p \geq 1$ integer:

$H^{-p}(0, 1)$:=

completion of $\mathcal{L}^2$ with the norm of the linear continuous functionals on $H_0^p(0, 1)$;

= extension to $H_0^p(0, 1)$ of the distributions of type

$$\sum_{k=0}^{p} f_k^{(k)}, \text{ with } f_k \in L^2(0, 1);$$

= extension to $H_0^p(0, 1)$ of the distributions of type

$$\sum_{k=0}^{p} (-1)^k f^{(2k)}, \text{ with } f \in H_0^p(0, 1).$$

From these choices it becomes quite clear that

$$\ldots \supset H^{-p} \supset \ldots \supset H^{-1} \supset H^0 = L^2 \supset H_0^1 \supset \ldots \supset H_0^p \supset \ldots, \qquad (4.54)$$

being each space densely and continuously contained in its immediate follower at the chain and being proper all the inclusions. The inclusion of H_0^p in H^{-p} is compatible with Riesz canonical isomorphism, which is given exactly by $\sum_{k=0}^{p}(-1)^k d^{2k}/dx^{2k}$.

To finish, we point the identification

$$(H^{-p})^* = H_0^p, \; p \geq 0 \text{ integer},$$

and the **generalized Schwarz formula**

$$| < f|g >_{L^2} | \leq \|f\|_{H^{-p}}\|g\|_{H^p}, \; f \in L^2, g \in H^p.$$

The same construction may be repeated in order to introduce $H^{-p}(I\!R)$ or $H^{-p}(\Omega)$, where $\Omega \subset I\!R^n$ is a regular open set.

Exercise 4.13 Consider the operators

$$D^\alpha : H^p \to H^{p-|\alpha|}, \; |\alpha| \leq p,$$

so as to generalize Exercise 4.8.

Exercise 4.14 Let $\psi \in L^2(0,1)$ and $-\phi'' + \phi = \psi, \phi \in H^1$. Then ϕ gives Riesz representation for the form L_ψ:

$$< L_\psi, f >:=< \psi|f >_{L^2}=< \phi|f >_{H^1}.$$

In particular, taking $\psi \in H^1$, we can have a look at how the same function ψ may be identified to different elements from H^{-1}. (For the case of $\tilde{H}^1$, take the boundary condition $\psi'(0) = \psi'(1) = 0$.)

4.7.4 The Spaces H^s, s Arbitrary Real

Let us recall the definition: a function u is said to be **locally Hölder-continuous** of order α, $0 < \alpha < 1$, **at** x_0, if

$$\sup_y \frac{|u(x_0 + y) - u(x_0)|}{|y|^\alpha} = K(x_0) < \infty.$$

And it owns the stronger characteristic of being **uniformly** Hölder-continuous of order α if

$$\sup_{x_0}\sup_{y}\frac{|u(x_0+y)-u(x_0)|}{|y|^{\alpha}}=K<\infty.$$

Such a property is weaker than differentiability but stronger than continuity. The constant K is a semi-norm for u in the space of order α uniformly Hölder-continuous functions.

It would be suitable to measure this property of u rather globally, not locally. Such motivation amounts to introduce, for $u:\mathbb{R}\to\mathbb{R}$,

$$|u|_{\alpha}^{2}:=\int\frac{dy}{|y|}\int\left[\frac{u(x+y)-u(x)}{|y|^{\alpha}}\right]^{2}dx. \tag{4.55}$$

The functions u for which $|u|^{\alpha}<\infty$ correspond to Hölder-continuous functions. This can be thought as the same trend that guided to the extension of the derivative concept within the framework of the order m Sobolev spaces, for positive integer m.

Define then H^s, $0<s<1$, as

$$H^{s}(\mathbb{R}):=\{f\in L^{2};\,|f|_{s}<\infty\},$$

with the norm

$$\|f\|_{H^{s}}^{2}:=\|f\|_{L^{2}}^{2}+|f|_{s}^{2}.$$

Observe that a given $f\in H^{p}$ belongs to the space H^{p+1} if $f^{(p)}\in H^{1}$, and

$$\|f\|_{H^{p+1}}^{2}=\|f\|_{H^{p}}^{2}+|f^{(p)}|_{1}^{2},$$

where $|\cdot|_1$ denotes the semi-norm $|u|_1:=\|u'\|_{L^2}$. This way define, for every $s\in\mathbb{R}$, $1\le m<s<m+1$,

$$H^{s}(\mathbb{R}):=\{f\in H^{m};\,f^{(m)}\in H^{s-m}\},$$

then introduce on it the norm

$$\|f\|_{H^{s}}^{2}:=\|f\|_{H^{m}}^{2}+|f^{(m)}|_{s-m}^{2}.$$

As long as this definition is at hand, introduce the spaces H^{-s} in a track similar to the one previously displayed for any positive integer$s$, and it is possible to complete the chain (4.54)

$$H^s \supset H^p \text{ if } s < p, s, p \in I\!R, \tag{4.56}$$

being each inclusion proper, continuous, and dense.

For $I\!R^n$, besides the notation to become a little heavier, it is further needed to replace in (4.55) $dy/|y|$ by $dy/|y|^n$. Lastly, when $\Omega \subset I\!R^n$ is a regular open set,

$$H^s(\Omega) := \{f \in L^2(\Omega); \exists \tilde{f} \in H^s(I\!R^n) \text{ which } \mathbf{extends} f\},$$

or, getting hold of the quotient spaces,

$$H^s(\Omega) := H^s(I\!R^n)/\{v; \text{supp } v \subset \Omega^c\}.$$

We close these short words on this topic with the mention of an alternate, which in several frameworks may turn out to be more operational. It consists in introducing these spaces H^s via

$$H^s(I\!R^n) := \{T \in \mathcal{S}; (1 + |\xi|^2)^{s/2}\hat{T} \in L^2\}.$$

In short, with the tempered distributions – described in the section that follows – whose Fourier transform has a special decay, cf. [1].

And still another possibility is to introduce the spaces H^s via interpolation theory, *à la Lions*, cf. [54].

4.8 Fourier and Laplace Transforms

A few words on the mathematical sentence: "Let the function f be known on the space $\mathcal{S}$." What are exactly the hypotheses carried over by these words? Usually they are understood as to assure our knowledge of some data associated to f, namely: its domain, its range – or at least an address where it may be found – and somehow an account of how x, in the domain, and $f(x)$ are related. Indeed, it is largely spread the feeling that all is needed so as to consider a studied function **known** is $x \to f(x)$. An explicit expression for it usually would give the feeling of work done, despite being overlooked questions as instability for *bona fide* computing and so on.

The analysis of some operators quite often aims to readily generate information about **intrinsic** properties from elements under study in the space where these operators act. Among these, it is worth mentioning derivatives, integrals, the curvature, the trace, determinants, and some other less familiar.

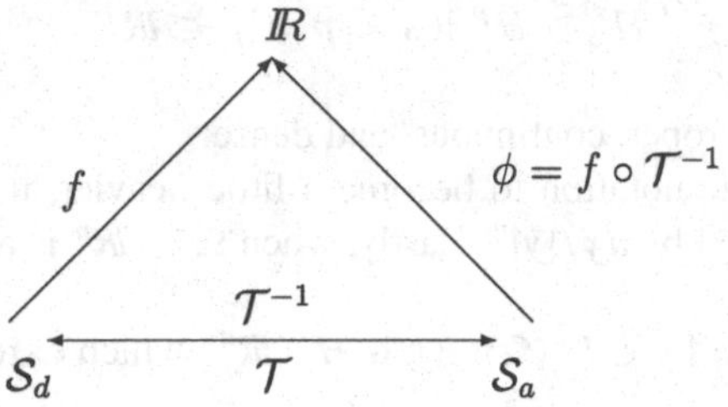

Other operators, baptized as **transforms**, have been created to, as we dwell on the correspondence between the elements and their images – either in the same space or in different ones – to make easier the access to searched information. We attain such results, then, from data collected on the arrival space $\mathcal{S}_a$. They let deducing properties we had been looking for in the initially working space, $\mathcal{S}_d$, the depart space. An **indirect** reading,[12] therefore.

4.8.1 Fourier Transform for Functions

It turns out that to operate with the Fourier transform in $L^1(I\!R)$ is a more appropriate choice than that one described in (2.33). As soon as $\tilde{L}^1(I\!R)$ is a dense subspace of $L^1(I\!R)$ and being $\mathcal{F}$ bounded, we are able to continuously extend it, in a unique way:

$$\left.\begin{array}{ll}\mathcal{F}: L^1(I\!R) \to C_a(I\!R) & \\ \quad f \quad \to (\mathcal{F}f)(t) := \int_{-\infty}^{\infty} f(x)e^{-itx}dx\end{array}\right] . \tag{4.57}$$

In spite of that, for a large amount of applications, this is not yet the most convenient framework to deal with. As a matter of fact, we have already pointed up the discontinuity of the inverse $\mathcal{F}^{-1}$, with respect to the norm $\|\bullet\|_\infty$ in $\mathcal{E}_i = C_a(I\!R)$. Besides, the functions on the range of $\mathcal{F}$, despite being more regular than those from $L^1(I\!R)$, since they are always continuous, may lack to be integrable on the real line, as taught by the

Exercise 4.15 Verify that $\mathcal{F}(\chi_A) \notin L^1(I\!R)$, for $A := [-a, a]$, $\forall a > 0$.

i. When $a = 1$, we have

$$\mathcal{F}(\chi_A) = 2\frac{\sin t}{t};$$

ii. For $0 \neq a \in I\!R$, if $h_a f(x) := f(ax)$, then

[12] Mathematicians prefer to call that an **inverse problem**.

$$\mathcal{F}(h_a f) = \frac{1}{a} h_{1/a} \mathcal{F}(f). \tag{4.58}$$

Another troublesome point when treating the Fourier transform with $L^1(\mathbb{R})$ chosen as the starting set is: How to characterize its range $\mathcal{E}_i := \mathcal{F}(L^1(\mathbb{R}))$? According to Example 5.8, we fail to deal with an onto mapping.

Consider in the domain of $\mathcal{F}$ functions ϕ with a stronger regularity, those whose derivatives belong to $L^1(\mathbb{R})$. These derivatives also have a Fourier transform, and this way we deduce, by using twice the Riemann-Lebesgue lemma, an expression for ϕ decay, cf. notation in (7.1):

$$\int \phi'(x)e^{-ixt}dx = -\int \phi(x)e^{-ixt}[-it]dx = i/[1/t]\mathcal{F}(\phi)(t)$$
$$\Longrightarrow \mathcal{F}(\phi)(t) = \mathbf{o}(1/|t|) \text{ for } |t| \to \infty.$$

With repeated use of integration by parts, we reach, for $n \geq 0$,

$$\phi^{[n]} \in L^1(\mathbb{R}) \Longrightarrow \mathcal{F}(\phi)(t) = \mathbf{o}(1/|t|^n) \text{ for } |t| \to \infty. \tag{4.59}$$

Theorem 2.2 from 2.14.7 allows to differentiate:

$$\frac{d}{dt}\mathcal{F}(\phi)(t) = \int \frac{d}{dt}\left[\phi(x)e^{-ixt}\right]dx = \int [-ix]\phi(x)e^{-ixt}dx. \tag{4.60}$$

Therefore, provided the first moment of ϕ exists, *i.e.*, if $x\phi(x)$ belongs to $L^1(\mathbb{R})$, we conclude to be differentiable its Fourier transform. In a similar way, we can deduce results that show a relationship between moments and higher-order derivatives.

Further, we have reached, under the hypotheses for (4.59) and (4.60), expressions that indicate how the operator $\mathcal{F}$ commutes with the differentiation operator. As long as we denote

$$(\nu_n f)(t) := (it)^n f(t) \quad \text{and} \quad (\mu_n \phi)(x) := (x/i)^n \phi(x),$$

they are

$$\mathcal{F}\left(\phi^{[n]}\right)(t) = (\nu_n \mathcal{F}(\phi))(t) \text{ and } (\mathcal{F}(\phi))^{[n]}(t) = \mathcal{F}(\mu_n \phi)(t). \tag{4.61}$$

The **support** of a real (or vector) function is defined as the closure of the complement of the set of all its zeroes, *i.e.*,

$$\operatorname{supp} f := \overline{f^{-1}(\{0\}^c)}.$$

Exercise 4.16 From an alternate view point, the support complement for a given function is the largest open set where it does not vanish. Therefore, if it does not vanish in a given point, such a point must belong to the support – which may as well also contain zeroes.

Consider the functions f that happen to be continuous and have a compact support. They form the subset $C_0^\infty(\mathbb{R})$ of the space $C_a(\mathbb{R})$ introduced in Example 2.4 on Sect. 2.12. For such functions f, all of their moments exist and therefore $\mathcal{F}(f) \in C^\infty(\mathbb{R})$. The reasoning just presented may lead one to wonder whether the best framework where to introduce the Fourier transform would be $C_0^\infty(\mathbb{R})$. Such a choice would be even better as long as it could be assured the support compactness for the transform of every function from $C_0^\infty(\mathbb{R})$. Unluckily we observe that, as a consequence of the above presented Exercise 4.15:

Given a compactly supported function f, to linearly expand or shrink its support – that is, to perform a **homothety** – induces a corresponding transformation in the norm $\|\mathcal{F}(f)\|_\infty$.

This property may be thought of as a clue to the existence of a relation between the support of f and the growth of $\mathcal{F}(f)$. The result that follows[13] points to this direction and even to look for another background where further deepening the search for $\mathcal{F}$.

Theorem (Paley-Wiener) *Let $f \in C_0^\infty(\mathbb{R})$ and assume for its support:*

$supp(f) \subset [-a, a]$.

It may then be deduced its transform $F := \mathcal{F}(f)$ to be analytic through $\mathbb{C}$ and that, to each integer N, to be associated a constant C_N in such a way that the inequalities below hold:

$$\frac{|F(\xi)|}{\exp(a|\xi_2|)} \leq \frac{C_N}{(1+|\xi|)^N}, \xi = \xi_1 + i\xi_2 \in \mathbb{C}. \tag{4.62}$$

Conversely, being F an analytic function in $\mathbb{C}$ which fulfills a chain of estimates like (4.62), then, for some $a > 0$, there exists in $C_0^\infty(\mathbb{R})$ a function f with

$supp(f) \subset [-a, a]$

and for which

$$F(\xi) = \mathcal{F}(f)(\xi).$$

Being analytic, the transform of a given function would have a compact support only if it is identically null, thus the need to search for another space to be mapped by $\mathcal{F}$ in itself. The choice goes to a space *close* to $C_0^\infty(\mathbb{R})$, which is the Schwartz space $\mathcal{S}$. This choice was pushed exactly by (4.62).

Before getting further, observe that the familiar Gaussian function is steadily seen to be an element of the space $\mathcal{S}$ and that for it holds the relation

[13] See [72] or [62].

$g(x) := \exp(-x^2/2)$

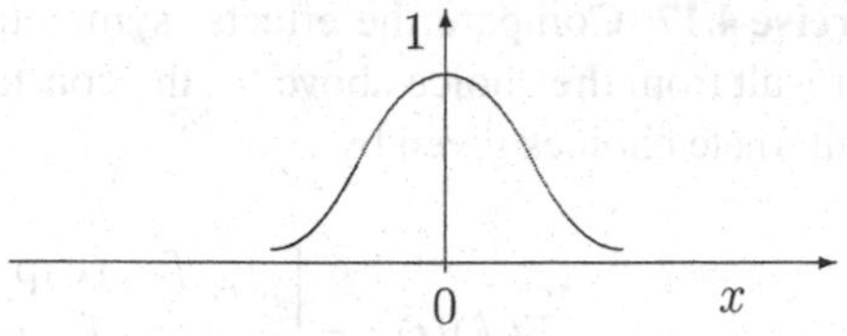

$$g'(x) = -xg(x), \forall x \in I\!R. \tag{4.63}$$

Apply $\mathcal{F}$ to both sides of (4.63) so as to reach

$$it\mathcal{F}(g)(t) = -i\mathcal{F}(\mu_1\phi)(t) = -i\frac{d}{dt}(\mathcal{F}(g))(t);$$

in other words, the same relation (4.63), now expressed for

$$G(t) := (\mathcal{F}(g))(t),$$

which gives

$$G'(t) = -tG(t).$$

From this conclusion, and due to the fact that g and G both live on the same domain, besides being g a never vanishing function, it follows:

$$\frac{d}{ds}G/g(s) = \frac{[-sG(s)]g(s) - G(s)[-sg(s)]}{g^2(s)} = 0.$$

As a consequence, there exists a constant λ for which

$$G(s) = \lambda g(s), \forall s \in I\!R.$$

Since

$$G(0) = \int g(x)dx = \sqrt{2\pi},$$

the choice of

$$\mathcal{F}(f)(t) := \sqrt{\frac{1}{2\pi}}\int_{I\!R} f(x)\exp(-itx)dx, \tag{4.64}$$

as a new definition assures the Gaussian function g to be a fixed point for $\mathcal{F}$.

Exercise 4.17 Compare the effects (symmetry, presence in the calculations, etc.) that result from the choice above for the constant in defining $\mathcal{F}$ with the ones led by the alternate choices given by

$$\mathcal{F}(f)(t) := \begin{bmatrix} \int_{\mathbb{R}} f(x)\exp(-2\pi itx)dx \\ (1/2\pi)\int_{\mathbb{R}} f(x)\exp(-itx)dx \\ \int_{\mathbb{R}} f(x)\exp(-\sqrt{2\pi}itx)dx \end{bmatrix},$$

besides, of course, the one in (2.33).

The space $\mathcal{S}$ of rapidly decreasing functions is mapped by $\mathcal{F}$ to a subspace of functions that also live in $\mathcal{S}$. This is a consequence of the above mentioned properties about decay both on the domain and on the image of $\mathcal{F}$. The equality $\mathcal{S} = \mathcal{F}(\mathcal{S})$ comes from the **inversion formula**,[14] valid for every $f \in \mathcal{S}$:

$$f(x) = \frac{1}{\sqrt{(2\pi)}}\int_{\mathbb{R}} F(t)\exp(ixt)dt,\ F := \mathcal{F}(f). \tag{4.65}$$

Thus,

$$f = \mathcal{F}(\rho F), \forall f \in \mathcal{S}, \text{ where } \rho F(t) := F(-t).$$

Moreover, from (4.65) follows

$$< f|g >_{L^2(\mathbb{R})} = < \mathcal{F}(f)|\mathcal{F}(g) >_{L^2(\mathbb{R})}$$

and now, to close, Plancherel-Parseval identity is presented:

$$\|f\|_2 = \|\mathcal{F}(f)\|_2, \forall f \in \mathcal{S}. \tag{4.66}$$

Taking into account that $\mathcal{S}$ is a dense subspace of $L^2(\mathbb{R})$, there exists a unique continuous extension for $\mathcal{F}$ to this space, so that we get hold of

Theorem 4.7 *The Fourier transform*

$$\mathcal{F} : L^2(\mathbb{R}) \to L^2(\mathbb{R}),$$

continuous extension for the operator defined by (4.64) for functions from $\mathcal{S}$, is an isometric isomorphism. Its inverse mapping is given, for regular elements from $L^2(\mathbb{R})$, by

$$\bar{\mathcal{F}}(f)(x) := \frac{1}{\sqrt{(2\pi)}}\int_{\mathbb{R}} f(t)\exp(ixt)dt. \tag{4.67}$$

[14] Cf. [40], pp.196, for an elegant proof.

Exercise 4.18 Observe that, for $(\tau^a f)(x) := f(x-a)$, it follows that

$$\mathcal{F}(\tau^a f)(t) = \exp(-iat)\mathcal{F}(f)(t)$$

and, therefore, it suffices to deduce (4.65) for $x = 0$ and for any f, to conclude the validity of the inversion formula.

Exercise 4.19 Examine the inherited properties by the restrictions

$$\mathcal{F} : H^n(I\!R) \to H^n(I\!R).$$

Let the subsection be closed by remarking that the just presented theoretical development remains valid for any space $I\!R^n, n > 1$, as long as all steps are duly adapted.

4.8.2 *The Tempered Distributions*

Section 4.4 introduces in the space of distributions $\mathcal{D}'(I\!R)$ operators, like translation and the derivative, that (initially) make sense only for spaces of *bona fide* functions. They all are examples of the so-called **definition by duality**: an operator Γ is defined in the dual space $\mathcal{D}'(I\!R)$ by means of its very **action** – otherwise said, through the action of an operator γ we seek to extend[15] – on the space $\mathcal{D}(I\!R)$,

$$\left.\begin{array}{l} \Gamma : \mathcal{D}' \to \mathcal{D}' \\ \qquad T \to \Gamma T : \mathcal{D} \to \qquad I\!R \\ \qquad\qquad f \to < \Gamma T, f >:=< T, \gamma f > \end{array}\right] .$$

The Fourier transform for distributions gets extended by this same track. But as long as the domain of $\mathcal{F}$ is the space $\mathcal{S}(I\!R) \supset \mathcal{D}(I\!R)$, a definition is on need for $\mathcal{S}'(I\!R)$, the topological dual for $\mathcal{S}(I\!R)$. But, beforehand, we must know which notion of convergence to deal with in $\mathcal{S}$ – and, consequently, in $\mathcal{S}'$. Once these constraints are established, the Fourier transform extension to the dual $\mathcal{S}'(I\!R)$ is straightened as

$$\left.\begin{array}{l} \mathcal{F} : \mathcal{S}' \to \mathcal{S}' \\ \qquad T \to \mathcal{F}T : \mathcal{S} \to \qquad I\!R \\ \qquad\qquad f \to < \mathcal{F}T, f >:=< T, \mathcal{F}f > \end{array}\right] . \tag{4.68}$$

[15] Still, more precisely, to generalize.

Being $\mathcal{F}$ linear in $\mathcal{S}$, it follows that its *extension* to $\mathcal{S}'$ is alike linear. As long as we work on[16] $\mathcal{S}'(I\!R)$ with the notion of continuity used in $\mathcal{D}'$, namely, the one associated to pointwise convergence, or weak* – cf. Sect. 6.3 –

$$T_J \stackrel{\mathcal{S}'}{\to} T \Longleftrightarrow < T_J, f > \stackrel{J}{\to} < T, f >, \forall f \in \mathcal{S},$$

if $\mathcal{F}$ is continuous in $\mathcal{S}$, the definition (4.68) will make it continuous in $\mathcal{S}'$.

We say that a sequence of functions (ϕ_J) in $\mathcal{S}$ is convergent, in the sense of the topology of $\mathcal{S}$, and ϕ is its limit if, for every $\psi_J := \phi_J - \phi \in \mathcal{S}$,

$$M(\psi_J, k, p) := \max_{x \in I\!R} \left| x^p \left(\frac{d^k \psi_J(x)}{dx^k} \right) \right| \stackrel{J}{\to} 0, \tag{4.69}$$

for arbitrary integers $k, p \geq 0$, cf. (2.1). In other words, if the sequence $(\psi_J) \in \mathcal{S}$ and, besides, the product of any of its derivatives with any polynomial converges uniformly to zero.

Since $\mathcal{S}(I\!R) \supset \mathcal{D}(I\!R)$, the distributions from $\mathcal{S}'(I\!R)$ compose a proper subset of $\mathcal{D}'(I\!R)$, as may be verified by considering, for example, the distribution associated to the locally integrable function $\exp(x^2)$. The elements of $\mathcal{S}'(I\!R)$ can not present such growing behavior, and that is the reason to be them called **tempered** distributions.

These distributions are characterized by the values they assume on $\mathcal{D}(I\!R)$, due to being this space dense in $\mathcal{S}(I\!R)$, as shown, say, with the

Exercise 4.20 Taken the function $\theta \in \mathcal{D}(I\!R)$ recalled in Exercise 2.17, the sequence (θ_n) in $\mathcal{D}(I\!R)$ defined by

$$(\theta_n)(x) := \begin{bmatrix} \theta(x/n) & |x| \leq n \\ \theta(x-n+1) & |x| \geq n \end{bmatrix},$$

fulfills $\theta_n \phi \stackrel{\mathcal{S}}{\to} \phi, \forall \phi \in \mathcal{S}$.

Worth to recall: the receipt to get $\mathcal{D}'(I\!R^n)$ is also fit to any open domain $\Omega \subset I\!R^n$, but the theory of tempered distributions is restricted to $I\!R^n$.

A distribution $T \in \mathcal{D}'(I\!R)$ is said to **vanish** on the open set $\mathcal{O} \subset I\!R$ if, for every $\phi \in C_0^\infty(I\!R)$, null on the complement of $\mathcal{O}$, we have $< T, \phi >= 0$. The **support** of T is then defined as

$$\text{supp } T := \bigcap \{\mathcal{O}^c | T \text{ vanishes on } \mathcal{O}\}.$$

[16] Again, our writing restricts itself to the line $I\!R$, but all concepts and results may be applied to $I\!R^n, n \geq 2$.

The constant function $f \equiv 1$, which we shall denote by $\mathbb{1}$, is associated to – or *is* – a tempered distribution. Let us obtain its Fourier transform:

$$< \mathcal{F}\mathbb{1}, \phi >:=< \mathbb{1}, \mathcal{F}\phi >= \int 1(\mathcal{F}\phi)(t)dt$$

$$= \int e^{i0t}\left[\frac{1}{\sqrt{2\pi}}\int e^{-ixt}\phi(x)dx\right]dt = \sqrt{2\pi}\phi(0).$$

This way we conclude that

$$\mathcal{F}\mathbb{1} = \sqrt{2\pi}\delta_0.$$

This result suggests to question whether it holds in $\mathcal{S}'(\mathbb{R})$ a relation that corresponds to (4.65), so that we could formulate in $\mathcal{S}'$ a similar result to the Theorem 4.7. The operator $\bar{\mathcal{F}}$, defined in (4.67), is introduced in $\mathcal{S}'$, in an analogous pattern to that one for $\mathcal{F}$. For the obtained operator, it holds

$$< \bar{\mathcal{F}}(\mathcal{F}T), \phi > = < (\mathcal{F}T), \bar{\mathcal{F}}\phi >$$
$$=< T, \mathcal{F}(\bar{\mathcal{F}}\phi) > = < T, \phi >, \forall\phi \in \mathcal{S}(\mathbb{R}).$$

Once the corresponding identity also shows to be valid for $\mathcal{F} \cdot \bar{\mathcal{F}}$, we are allowed to announce the

Theorem 4.8 *The Fourier transform* $\mathcal{F} : \mathcal{S}'(\mathbb{R}) \to \mathcal{S}'(\mathbb{R})$ *is a linear continuous operator, besides being 1-1 and onto. Its inverse operator happens to be also continuous and further given by* $\bar{\mathcal{F}}$.

Observe for the distribution $\mathbb{1}$ that the support of its Fourier transform equals $[-a, a]$, with $a = 0$, and thus the estimate $|\mathbb{1}(x)| \le e^{ax}$ holds. This fact suggests to search for an extension of the Paley-Wiener Theorem for distributions. That is what the result that follows exposes, cf. [72].

Theorem (Paley-Wiener for Distributions) *Let* $T \in \mathcal{S}(\mathbb{R})$. *Its Fourier transform* $F := \mathcal{F}(T)$ *is an entire function – i.e., analytic in all of* $\mathbb{C}$ *– besides, there exist constants* $N \in \mathbb{N}$, $C > 0$, $a > 0$ *for which:*

$$\frac{|F(\xi)|}{\exp(a|\xi_2|)} \le C(1 + |\xi|)^N, \xi = \xi_1 + i\xi_2 \in \mathbb{C}. \tag{4.70}$$

Conversely, if F is analytic on $\mathbb{C}$ and fulfills estimates like (4.70), for positive constants a, C, and $N \in \mathbb{N}$, then there exists a distribution $T \in \mathcal{S}(\mathbb{R})$ such that $F := \mathcal{F}(T)$.

Another characterization of the tempered distributions is contained in the following

Theorem 4.9 *To any distribution $T \in \mathcal{S}(\mathbb{R})$, there corresponds an integer ℓ and a function $f \in C^0(\mathbb{R})$ with a polynomial growth – i.e., such that there exists a polynomial which dominates $f(x)$ on its whole domain, $|f(x)| \leq |p(x)|$ – and besides such f satisfies*

$$T = D^\ell f.$$

The derivative in the sense of $\mathcal{D}$ is generally denoted by D. And the converse of Theorem 4.9 also holds.

To close the present subsection, let us put some emphasis on the flexibility that the distributions have handed to the Fourier transform, as it has opened alternate tracks to the treatment of differential equations.

4.8.3 Laplace Transform

Consider a function $f : \mathbb{R} \to \mathbb{C}$, null for $\{t<0\}$ as well as locally integrable. Given $p \in \mathbb{C}$, $p := \xi + i\eta = \Re(p) + i\Im(p)$, for which the integral

$$\int_{\mathbb{R}} f(t)e^{-pt}dt = \int_0^\infty f(t)e^{-pt}dt \tag{4.71}$$

exists and is finite, its value, denoted by $\mathcal{L}[f](p)$, defines the **Laplace integral** of f at the point p.

Observe that, given f, the existence of $\mathcal{L}[f](p)$ depends only on the real part of p, which suggests:

The **summability abscissa** $a = a_f$ of the Laplace integral of a function f is defined as

$$a := \inf\{\xi \in \mathbb{R} | \int_{\mathbb{R}} e^{-\xi t}|f(t)|dt < \infty\}.$$

Exercise 4.21 Being H the Heaviside function,

(i) For $f_{\mp}(t) := \exp(\mp t^2)H(t)$, we have $a_{f\mp} = \mp\infty$.
(ii) For

$$\left.\begin{array}{l} h(t) := H(t-1)/t^2,\ g(t) := H(t)e^t, \\ a_h = 0,\ \mathcal{L}[h](0) = 1,\ a_g = 1,\ \int e^{-t}g(t)dt = +\infty \end{array}\right],$$

and thus, in general, the integral $\mathcal{L}[f](a_f)$ fails to be defined. •

Given a function f for which $a_f < +\infty$, it is possible to define its **Laplace transform** on the half plane $\{p \in \mathbb{C} | \Re(p) > a_f\}$ as

$$\mathcal{L}[f](p) := \int_0^\infty f(t)e^{-pt}dt. \tag{4.72}$$

This transform mimics then the Fourier transform as regards to an important relation with the differentiation operator, namely, the latter is mapped on a much simpler operator. Besides, an image function gotten by Laplace transform shows up much more regularity than those given as images by Fourier transform. These are the facts described by the following:

Proposition *Suppose that f has $a \in \mathbb{R}$ as its summability abscissa. Then $\mathcal{L}[f](p)$ is* ***holomorphic*** *on its domain, the half plane*

$$\mathcal{H}_p := \{p \in \mathbb{C} | \Re(p) > a\}.$$

For $m \geq 0$, the functions $[(-t)^m f(t)]$ have all the same summability abscissa a and, denoting the derivatives in $\mathbb{C}$ by $D^m := d^m/dp^m$, the following relations hold:

$$D^m \mathcal{L}[f](p) = \mathcal{L}[(-t)^m f](p), m \geq 1. \tag{4.73}$$

Therefore, the operator $\mathcal{L}$ maps a set of functions into another, having each image function eventually a different domain, which depends on the corresponding origin function. This is the reason why the distributions for Laplace transform fail to be introduced throughout the now standard definition via duality. The adopted path may be thought of as a more straight generalization for (4.71): since the functions on the domain of $\mathcal{L}$ *are* distributions,[17] we write

$$\mathcal{L}[f](p) =< T_f, \phi_p >, \text{ with } \phi_p(t) := e^{-pt},$$

and struggle to formalize

$$\mathcal{L}[T](p) :=< T, \phi_p >, \text{ with } T = \text{[some distribution]}. \tag{4.74}$$

We have not even specified which is the distribution space $\mathcal{L}$ operates on. Worse than that, how to attribute a meaning to the right hand side of (4.74), since ϕ_p fails to be a test function, at least not for the spaces we are dealing with, $\mathcal{D}$ and $\mathcal{S}$?

Let us denote by $\mathcal{D}'_+$ the space of the distributions whose support is contained in $\mathbb{R}_+ := [0, +\infty[$ and let us adapt to this space the construction of the Laplace transform for functions, taking, by $T \in \mathcal{D}'_+$:

$$\sigma(T) := \inf\{\xi_0 \in \mathbb{R} | \phi_{-\xi} T \in \mathcal{S}, \forall \xi > \xi_0\}. \tag{4.75}$$

[17] Throughout this subsection the functions from $\mathcal{D}, \mathcal{S}$ are considered with values on $\mathbb{C}$; the distributions from $\mathcal{D}', \mathcal{S}$ are thus complex functionals.

In the subspace of $\mathcal{D}'_+$ given by

$$\mathcal{D}_L := \{T \in \mathcal{D}'_+ | \sigma(T) < +\infty\},$$

the definition (4.74) is put into a formal status as follows.

Take $\alpha \in C^\infty(\mathbb{R})$ which satisfies

$$\alpha(x) = \begin{bmatrix} 0 & x \in]-\infty, x_0] \\ 1 & x \in [0, +\infty[\end{bmatrix},$$

being arbitrary $x_0 < 0$ as well as the values assumed by α on $]x_0, 0[$.

Given $T \in \mathcal{D}_L$, let $p \in \mathbb{C}$ with $\xi := \Re(p) > \sigma(T)$ and $\xi_1 \in]\sigma(T), \xi[$, in such a way as we then have

$$\alpha(t)\phi_{-(p-\xi_1)}(t) = \alpha(t)e^{-(p-\xi_1)t} \in \mathcal{S}, \phi_{-\xi_1}(t)T = e^{-\xi_1 t}T \in \mathcal{S}'.$$

This will allow to define

$$\mathcal{L}[T](p) :=< \phi_{-\xi_1}T, \alpha\phi_{-(p-\xi_1)} > . \tag{4.76}$$

It is then seen that, fixed $T \in \mathcal{D}_L$ and $p \in \{z \in \mathbb{C} | \Re(z) > \sigma(T)\}$, the value assumed by the expression (4.76) is independent of α and of ξ_1. Therefore, we are assured of the consistency of the proposed definition for $\mathcal{L}[T](p)$. We then informally write its value as

$$\mathcal{L}[T](p) :=< T, e^{-pt} > .$$

Exercise 4.22 Prove the **linearity** of $\mathcal{L}$, in a sense to be made precise.

The operator $\mathcal{L}$ exhibits the following additional properties:

1-1*ness* Given $U, T \in \mathcal{D}_L$,

$$\mathcal{L}[U] = \mathcal{L}[T] \Longrightarrow U = T.$$

Polynomial growth The transform $\mathcal{L}[T]$ of each distribution $T \in \mathcal{D}_L$ presents a polynomial growth, *i.e.*, for each closed half plane $R_c := \{z | \Re(z) \geq c\}$ contained in the domain of $\mathcal{L}[T]$, a polynomial P_c may be determined which fulfills

$$|\mathcal{L}[T](p)| \leq |P_c(p)|, \forall p \in R_c. \tag{4.77}$$

Conversely, given a holomorphic function $\Gamma(p)$, whose domain contains some half plane of type R_c, in order to be sure that it happens to be the Laplace transform of some distribution $T \in \mathcal{D}'_+$,

$$\Gamma(p) = \mathcal{L}[T](p),$$

it is enough that it shows a polynomial growth. In other words, there exists some polynomial and some half plane such that an estimate like (4.77) holds for them.

The proof of these properties demands strongly the tempered distributions representation assured by Theorem 4.9 on the previous section, cf. [8, 65].

Let us close the chapter bringing still another link between the two transforms just discussed. Observe that every distribution T of compact support is tempered. By the same route used to reach Laplace transform in $\mathcal{D}_L$, one may introduce – by using the same functions α in order to truncate the support of $\phi_{-itx} := \exp(-itx)$, $x \in I\!R$, but with no changes for its values on the support of T – the function:

$$< T, \phi_{-it} > := < T, \alpha\phi_{-it} > . \tag{4.78}$$

It may then be seen that such a function *is* the distribution $\mathcal{F}T$. Besides, it may as well be extended to the whole plane as a holomorphic function, and it fulfills (4.78) for $t \in \boldsymbol{C}$. In short, we can count on

Theorem 4.10 *The Fourier transform of a distribution of compact support is an entire function whose numerical values are given by (4.78).*

Chapter 5
The Three Basic Principles

5.1 Introduction

This chapter is devoted to the so-called three Functional Analysis basic principles: the Hahn-Banach Theorem about continuous extension of linear forms; the Open Mapping Theorem – one of whose consequences has been used in the proof of Theorem 5.1, Sect. 3.2. and the Banach-Steinhaus Theorem, or Uniform Boundedness Principle, which we have also mentioned previously (Theorem 4.2). As long as we will skip the proofs for all these three results, our aim through the sections that follow[1] is to motivate the readers, trying to convince them to be "reasonable" these statements, as well as pointing up how to take hold of them, by showing some of their relevant applications.

Rigorous proofs for these results may be found practically in any Functional Analysis book, particularly in [6, 39, 55, 60, 61, 67]. All of them discuss these results at the same level of generality as these notes do, while [69, 72] present a more general treatment.

5.2 Hahn-Banach Theorem

Many consequences follow from the characterization for dense subspaces in a normed space, as described in

Theorem 5.1 *A subset $D \subset N$ is dense if and only if, for every form $\ell \in N^*$,*

$$\ell(D) = \{0\} \Longrightarrow \ell \equiv 0.$$

---

[1] This chapter has taken the option to denote by N, N_1, N_2, etc., or B, B_1, B_2, ..., normed or Banach spaces, respectively.

C. A. de Moura, *Functional Analysis Tools for Practical Use in Sciences and Engineering*, https://doi.org/10.1007/978-3-031-10598-2_5

Proof As long as D is dense and ℓ continuous, ℓ must be null if it vanishes on D.

Conversely, assume that $\bar{D} \neq N$ and accept the following:

Proposition *Every closed proper subspace from N is contained in a closed hyperplane.*

Thus, $\bar{D}$ is then in a closed hyperplane H. Based on remarks from Sect. 3.2, it is possible to find a non-null functional $\ell_H \in N^*$ which has H as its kernel. Therefore, $\ell_H(D) = \{0\}$ and $\ell_H \neq 0$, but this contradicts the assumed hypothesis.

Theorem 5.1 proof was thus reached, except for the above stated Proposition, whose proof is missing. In some sense this proposition looks intuitive, nonetheless its proof is not straightforward. It follows precisely from Hahn-Banach Theorem.

For any Hilbert space, its dual has been completely characterized by Riesz Representation Theorem. On the other hand, in Sect. 3.5 some Banach spaces have had their dual characterizations presented: c_0, c, ℓ^p, and $L^p(\Omega), 1 \leq p < \infty$. In spite of that optimistic start, for an arbitrary normed espace N, based on the information we own up to this point, it is not even possible to claim that in N^* there exists some non-null functional. Indeed, whenever N has infinite dimension, the existence of $\ell \in N', \ell \neq 0$ may be only assured by employing the concept of a Hamel basis, cf. Sect. 8.2. Such concept also leads to the proof that provided $N' = N^*$, then $\dim(N) < \infty$. But, in order to get information about the existence of non-null elements in N^*, the required tool is the following:

Extension Theorem (Hahn-Banach) *Let $S \subset N$ be a subspace and ℓ a functional in S^*. In this case, there always exists $\tilde{\ell} \in N^*$ with*

$$\|\tilde{\ell}\|_{N^*} = \|\ell\|_{S^*}, \quad \tilde{\ell}|_S = \ell. \tag{5.1}$$

We call readers attention to the fact that this theorem does not require the **completeness** hypothesis for N, neither assumes to be S **closed**.

The proof main idea is quite simple: just reach a continuous extension ℓ_1 to ℓ for a subspace $S_1 := S \bigoplus [w]$, where $[w] := \{v := \alpha w; \alpha \in \mathbb{R}\}$ and $w \notin S$, in such a way that ℓ_1 and ℓ have the same norm. That is the tricky chunk of the proof. This step is then repeated for ℓ_1 and S_1. To deduce that such process leads to the extension $\tilde{\ell}$ claimed in the theorem statement, a transfinite argument is needed, the important **Zorn Lemma**.

Example 5.1 Given $w \in N$, there exists $\ell \in N^*$, with

$$\|\ell\| = 1, \quad \ell w = \|w\|. \tag{5.2}$$

Indeed, the linear functional

$$\begin{aligned} \ell_w : \quad [w] \quad &\to \mathbb{R} \\ v := \alpha w &\to \ell v := \alpha \|w\| \end{aligned} \tag{5.2a}$$

fulfills $\|\ell\|_{[w]^*} = 1$, which gives the claimed result. From it there follows the

Example 5.2 Given $v_1 \neq v_2$ in N, there exists $\ell \in N^*$, with

$$\ell v_1 \neq \ell v_2, \ \|\ell\| = 1.$$

Another direct consequence follows:

Example 5.3 Given $v \in N$, it can be claimed that

$$\|v\| := \sup_{\{\ell \in V^* \|\ell\|=1\}} \ell v = \sup_{\{\ell \in V^* \ell \neq 0\}} \frac{\ell v}{\|\ell\|}. \tag{5.3}$$

The relation (5.3) is known as **dual formula** (or dual variational formula) for the norm associated to v: get it compared to

$$\|\ell\| := \sup_{\{v \in V \|v\|=1\}} \ell v = \sup_{\{v \in V v \neq 0\}} \frac{\ell v}{\|v\|}. \tag{5.3*}$$

The identity (5.3) lets us to write, for example, for $1 \leq p < \infty$and conjugated exponent q

$$\|f\|_p = \sup_{\{g \in L^q(\Omega) \|g_q\|=1\}} \int_\Omega f(x)g(x)dx, \quad \forall f \in L^p(\Omega), \tag{5.3a}$$

$$\|\xi\|_p = \sup_{\{\eta=(\eta_j) \in \ell^q; \|\eta\|_q=1\}} \sum_{j=1}^{\infty} \xi_j \eta_j, \quad \forall \xi = (\xi_j) \in \ell^p. \tag{5.3b}$$

From (5.3) it may also be deduced that equality holds in (3.13), *i.e.*,

$$\|J_v\|_{V^{**}} = \|v\|_V,$$

for every $v \in V (\equiv$ normed space), being $J_\bullet$ the canonical identification between V and its bidual V^{**} introduced in Sect. 3.6.

Example 5.4 Given a closed subspace F and $w \notin F$, there exists $\ell \in N^*$, with $\ell(F) = \{0\}$ and $\ell w \neq 0$.

In other words, given a closed subspace F and a vector w outside it, there exists a closed hyperplane which contains F but not w. This is precisely what is told by the Proposition mentioned to get the proof of the above stated Theorem 5.1.

Exercise 5.1 Any closed subspace $F \subset N$ may be described as the intersection of all closed hyperplanes that contain itself.

Exercise 5.2 Prove the **conjugated variational principles** $(A) - (B)$ that follow.

Given the subspace $S \subset N$, let $F := \{\ell \in N^*;\ S \subset ker(\ell)\}$. Then it holds

$$(A) \qquad \operatorname{dist}(x, S) := \inf_{s \in S} \|x - s\| = \sup_{0 \neq \ell \in S} \ell x / \|\ell\|,$$

for every $x \in N$;

$$(B) \qquad \operatorname{dist}(n, F) := \inf_{\ell \in F} \|n - \ell\| = \min_{\ell \in F} \|n - \ell\| = \sup_{0 \neq s \in S} ns / \|s\|,$$

for every $n \in N^*$.

A normed (or metric) space is said to be **separable** whenever it contains a dense countable subset. Exercise 2.15 shows that $C^0[0, 1]$ is separable. Other separable spaces are presented in

Example 5.5 All spaces ℓ^p, with $1 \le p < \infty$, are separable.

As a matter of fact, the set $\{q := (q_j) \in \ell^p;\ q_j \in \mathbb{Q},\ j \in \mathbb{N}\}$ is countable and dense.

Example 5.6 The space ℓ^∞ fails to be separable.

Indeed, suppose that $\{x^n\}$ is a countable and dense set in ℓ^∞, with $x^n := (x^n_j)$. The set $\{\hat{x}_n\}$ with $\hat{x}^j_n := x^j_n / \|x^n\|_\infty$ will then be dense[2] in the unity ball of ℓ^∞. Let us define $y = (y_j) \in \ell^\infty$ by

$$y_j := \begin{bmatrix} -\operatorname{sgn}(x^j_j) & x^j_j \neq 0 \\ 1 & x^j_j = 0 \end{bmatrix}.$$

It may be then deduced from this that $\|y - \hat{x}^n\|_\infty \ge 1, \forall n \in \mathbb{N}$, $\|y\|_\infty = 1$. But this claim contradicts to be $\{\hat{x}_n\}$ dense in the unity ball.

Now we shall make use of Theorem 5.1 to prove that $(\ell^\infty)^* \neq \ell^1$, in the sense of Exercise 3.4, or more precisely, on Theorem 3.3 from Sect. 3.5. At first we need to make use of

Theorem 5.2 *When N^* is separable, N is separable as well.*

Proof Let $\{\phi_n\}$ be a countable dense set in the unitary ball from N^*. It can be claimed that the set $\{x^n\}$ is dense in the unity ball of N, where x^n is chosen in such a way as to fulfill

[2] It may be assumed – with no loss of generality – that the null sequence is out of this set.

$$|\phi_n x^n| \geq \|\phi_n\|/2, \quad \|x^n\| \leq 1.$$

If $\{x^n\}$ fails to be dense, by Theorem 5.1 it is possible to determine a functional $\phi \in N^*$, with $\|\phi\| = 1$, such that $\phi x^n = 0, \forall n \in \mathbb{N}$. Since $\{\phi_n\}$ is a dense set in the unity ball of N^*, there exists a sequence $\phi_{n_k} \to \phi$. From this it is deduced that

$$\|\phi - \phi_{n_k}\| \geq |(\phi - \phi_{n_k})x^{n_k}| = |\phi_{n_k} x^{n_k}| \geq \|\phi_{n_k}\|/2.$$

Since $\|\phi - \phi_{n_k}\| \to 0$, it follows that $\phi_{n_k} \to 0$, which gives $\phi \equiv 0$, and this contradicts to be $\|\phi\| = 1$.

Example 5.7 The space ℓ^∞ is not reflexive. In particular $(\ell^\infty)^* \neq \ell^1$.

In fact, had we $(\ell^\infty)^* = \ell^1$, then ℓ^∞ would be necessarily separable, by Theorem 5.2.

5.2.1 Application: A Dirichlet Problem

An application of the Hahn-Banach Theorem is described in the sequel. Its aim is to reach the solution, with the use of Green's function, for the following Dirichlet problem.

Let $\mathcal{D}$ be an open bounded and connected set in the plane xy, whose boundary $\mathcal{C}$ is made up of a finite number of regular curves. Besides let us suppose that each point p in $\mathcal{C}$ is one of the endpoints of a segment whose points, with the exception of p, are all in the exterior[3] of $\mathcal{D}$.

Given a continuous function f on $\mathcal{C}$, the aim is to determine on $\mathcal{D}$ a harmonic function u – *i.e.*, one that fulfills

$$\Delta u := \frac{\partial^2 u}{\partial x^2} + \frac{\partial^2 u}{\partial y^2} = 0-,$$

which is continuous on $\overline{\mathcal{D}}$ and coincides with f on $\mathcal{C}$; in short, $u \in C^0(\overline{\mathcal{D}})$ and is such that

$$\left[\begin{array}{l} \Delta u = 0 \quad \text{on } \mathcal{D} \\ \quad u = f \quad \text{in } \mathcal{C} \end{array}\right. . \tag{5.4}$$

Denoting by $\mathcal{H}$ the space of the continuous functions on $\overline{\mathcal{D}}$, harmonic on $\mathcal{D}$, and by $\mathcal{V}$ the space of the real continuous functions on $\mathcal{C}$, the maximum modulus principle for harmonic functions assures that the linear operator

[3] Exterior of $\mathcal{D} := \overline{\mathcal{D}}^c$ (= complement of the closure of $\mathcal{D}$).

$$A : \mathcal{H} \to \mathcal{V}$$
$$u \to Au := u|_C,$$

is 1-1. Therefore, we can count on the existence of the inverse operator

$$A^{-1} : Im(A) \subset \mathcal{V} \to \mathcal{H}$$

and thus all amounts to determine $\mathcal{V}_0 := Im(A)$, which means to know for which functions from $\mathcal{V}$ is it possible to solve this Dirichlet problem. Observe that as long as the norm $\|\cdot\|_\infty$ is taken in $\mathcal{H}$ as well as in $\mathcal{V}$, by the maximum modulus principle both operators A and A^{-1} turn out to be continuous.

On which follows we will need the **Green identities**: for $u, v \in C(\overline{\Omega})$, with $\Omega \subset \mathbb{R}^p$, it may be proven that

$$\int_\Omega [v\Delta u + (\nabla u|\nabla v)]\, dX = \int_{\partial\Omega} v(\nabla u|n) dS, \tag{5.5}$$

where

$$\nabla u := (\partial u/\partial x_1, \ldots, \partial u/\partial x_p)$$

is the **gradient** of u, $dX := dx_1 \ldots dx_p$ and n denotes the outer normal to the surface $\partial\Omega$, which is the boundary of Ω. The identity (5.5), called **first Green's formula**, is a consequence of the divergence (or Gauss) theorem: for an arbitrary vector field $F := (F_1, \ldots, F_p) \in C^1(\Omega)$, being

$$\mathrm{div} F := (\nabla|F) = \sum_{\iota=1}^{p} \partial F_\iota/\partial x_\iota$$

the **divergence** of F, then

$$\int \mathrm{div} F dX = \int_{\partial\Omega} (F|n) dS \tag{5.6}$$

holds. To have it checked it suffices to apply (5.6) to $F := v\nabla u$.

Now in (5.5) exchange u and v, then subtract the resulting identity from the original one, so as to be led to the so-called **second Green's formula**:

$$\int_\Omega [u\Delta v - v\Delta u]\, dX = \int_{\partial\Omega} \left[u\frac{\partial v}{\partial n} - v\frac{\partial u}{\partial n} \right] dS, \tag{5.7}$$

where $\partial u/\partial n := (\nabla u|n)$ denotes the u derivative in the direction of n. In particular, if u, v happen to be both harmonic, it follows that

$$\int_{\partial\Omega}\left[u\frac{\partial v}{\partial n}-v\frac{\partial u}{\partial n}\right]dS=0. \tag{5.7a}$$

By taking now $v := 1$, this conclusion is reached: if u is harmonic in Ω,

$$\int_{\Gamma}\frac{\partial u}{\partial n}\,dS=0, \tag{5.8}$$

for any surface Γ contained in Ω.

From (5.8) one may deduce the **mean value property** for harmonic functions. Take for a given $x_0 \in \Omega$ the surface

$$\Gamma := \{y \in \mathbb{R}^p;\ |y-x_0| \le r\} \subset \Omega.$$

Then

$$\begin{aligned}0&=\int_{|y-x_0|=r}\frac{\partial u(y)}{\partial n}dS=\int_{|y|=1}\left[\frac{d}{dr}u(x_0+ry)\right]r^{p-1}dS\\&=r^{p-1}\frac{d}{dr}\int_{|y|=1}[u(x_0+ry)]\,dS,\end{aligned}$$

which implies, for $0 < r \le r_0 = r_0(x_0)$, that

$$\int_{|y|=1}[u(x_0+ry)]\,dS=\alpha=\text{ constant}.$$

This constant will be obtained by employing the Mean Value Theorem from Integral Calculus linked to the continuity of u:

$$\alpha=\lim_{r\to 0}\int_{|y|=1}u(x_0+ry)dS=u(x_0)\omega_{p-1},$$

being ω_{p-1} the surface area of the unitary sphere in $\mathbb{R}^p$. This way, for arbitrary $x_0 \in \Omega$, we obtain

$$\begin{aligned}u(x_0)&=\frac{1}{\omega_{p-1}}\int_{|y|=1}u(x_0+ry)dS\\&=\frac{1}{\omega_{p-1}}\int_{|y-x_0|=r}u(z)\frac{dS}{r^{p-1}},\end{aligned}$$

otherwise said,

$$u(x_0) = \frac{1}{\omega_{p-1} r^{p-1}} \int_{|z-x_0|=r} u(z) dS. \tag{5.9}$$

Exercise 5.3 The expression (5.9) is called **first mean value property**. Prove its equivalence to

$$u(x_0) = \frac{p}{r^p \omega_p} \int_{|z-x_0|\leq r} u(z) dS, \tag{5.10}$$

known as **second mean value property**.

It can be verified to be a consequence of (5.9) the maximum modulus principle for harmonic functions.

Now let us get back to the case $p = 2$. We point up that the general context $p > 2$ may be read, for example, in [38] or [41]. Let

$$P_0 := (x_0, y_0) \in \mathcal{D} \text{ fixed}, \ r := \sqrt{[(x-x_0)^2 + (y-y_0)^2]} > 0$$

and

$$v(x, y) := \ln r.$$

As long as $\epsilon > 0$ is small enough, applying (5.7) to $\mathcal{D}_\epsilon := \mathcal{D} \backslash \mathcal{B}_\epsilon$, with $\mathcal{B}_\epsilon := B(P_0; \epsilon)$, it follows that

$$\begin{aligned} 0 &= \left\{ \int_C + \int_{\partial \mathcal{B}_\epsilon} \right\} \left[u \frac{\partial}{\partial n} \ln r - \ln r \frac{\partial u}{\partial n} \right] dS \\ &= \int_C \left[\frac{\partial}{\partial n} \ln r - \ln r \frac{\partial u}{\partial n} \right] dS - \ln \epsilon \int_{\partial \mathcal{B}_\epsilon} \frac{\partial u}{\partial n} dS \\ &\quad + \int_{\partial \mathcal{B}_\epsilon} u \frac{\partial}{\partial n} (\ln \epsilon) dS. \end{aligned}$$

Take (5.8) into account and then recall that on $\partial \mathcal{B}_\epsilon$ we have $\partial \ln \epsilon / \partial n = 1/\epsilon$, in order to deduce that

$$\frac{1}{\epsilon} \int_{\partial \mathcal{B}_\epsilon} u dS = \int_C \left[u \frac{\partial}{\partial n} \ln r - \ln r \frac{\partial u}{\partial n} \right] dS,$$

or alternatively, due to (5.9),

$$u(x_0, y_0) = \frac{1}{2\pi} \int_C \left[u \frac{\partial}{\partial n} \ln r - \ln r \frac{\partial u}{\partial n} \right] dS. \tag{5.11}$$

Subtract then (5.11) from (5.7a) to get

$$u(x_0, y_0) = \frac{1}{2\pi} \int_C \left[u \frac{\partial}{\partial n}(v - \ln r) - (v - \ln r) \frac{\partial u}{\partial n} \right] dS.$$

Suppose v to be harmonic on $\mathcal{D}$ and that it coincides with $\ln r$ in $\mathcal{C}$. Deduce then that

$$u(x_0, y_0) = \frac{1}{2\pi} \int_C u \frac{\partial}{\partial n}(v - \ln r) dS. \tag{5.12}$$

A function v for which these properties hold lets one to solve, from (5.12), the problem posed in (5.4) for an arbitrary f. This is which has motivated us to introduce the notion of the Green function $G(p; p_0)$ associated to the Dirichlet problem under discussion: it is a function which, being continuous in $\overline{\mathcal{D}}$, for each fixed $p_0 \in \mathcal{D}$ fulfills

$$\left.\begin{array}{ll} a. & G(p; p_0) = 0 \quad p \in \mathcal{C} \\ b. & \Delta G(p; p_0) = 0 \quad p \in \mathcal{D} \end{array}\right] . \tag{5.13}$$

Our aim now is to prove the existence of G.

For each fixed point $q \in \mathcal{D}$, the linear functional $l_q := \delta_q A^{-1}$ is continuous (here we are using the notation on Example 2.12b, Sect. 2.5). More precisely, $\|l\|_q = 1$, since for $f \equiv 1$, $A^{-1} f \equiv 1$. Hahn-Banach Theorem assures then the existence of an extension for l_q, denoted by $L_q \in V^*$, with $\|L_q\| = 1$.

Observe that, for each fixed $p \in \mathbb{R}^2 \backslash \mathcal{C}$, the function

$$s \mapsto g_p(s) := \ln |s - p|$$

holds continuity in $\mathcal{C}$, *i.e.*,

$$p \notin \overline{\mathcal{D}} \Longrightarrow g_p \in \mathcal{V}.$$

Since, for fixed $p \in \mathbb{R}^2$, the function $s \mapsto \ln |s - p|$ is harmonic on $\mathbb{R}^2 \backslash \{p\}$, we conclude that

$$p \notin \overline{\mathcal{D}} \Longrightarrow g_p \in \mathcal{V}_0.$$

Define then the Green function $G(p; q)$ for the Dirichlet problem under study: for fixed $p \in \mathbb{R}^2$, assign

$$G(p; q) = -\ln |q - p| + k(p, q), \tag{5.14}$$

where

$$k(p, q) = \begin{bmatrix} L_q(g_p) & p \notin \mathcal{C} \\ \ln|q-p| & p \in \mathcal{C} \end{bmatrix}.$$

Continuity for G ought to be proven, besides that (5.13b) holds, since (5.13a) is fulfilled, thanks to the introduced definition.

In fact, let fix$q$. For $p \in \mathcal{C}$, $\Delta k(p, q) = 0$, while for $p \notin \mathcal{C}$, we have

$$0 = L_q \Delta q_p = \Delta k(p, q).$$

The second equality must be justified, which amounts to verify that the operators Δ and L_q commute. With such purpose, introduce, for $h > 0$, the discretized operators

$$\Delta^h : s(x, y) \mapsto [s(x + hy) + s(x - hy) \\ -4s(x, y) + s(x, y + h) + s(x, y - h)]/h^2.$$

It is seen that, provided $v \in C^3(\mathcal{D})$, the identities

$$\lim_{h \to 0} \Delta^h v(x, y) = \Delta v(x, y)$$

hold uniformly in each compact of $\mathcal{D}$. Therefore, since

$$\Delta^h L_q g_p - L_q \Delta g_p = L_q(\Delta^h g_p - \Delta g_p),$$

the validity of

$$\|\Delta^h L_q g_p - L_q \Delta g_p\|_\infty \leq \|\Delta^h g_p - \Delta g_p\|_\infty \to 0$$

holds, as long as the sup in the norm $\| \cdot \|_\infty$ is taken on an arbitrary compact in $\mathcal{D}$. As a consequence, (5.13b) holds, which implies continuity for G on $\mathcal{D}$.

It rests on waiting the proof for the continuity of G for any $p \in \mathcal{C}$.

When $p \notin \overline{\mathcal{D}}$ and $q \in \mathcal{D}$, then $L_q g_p = \ln|p - q|$. Thus, for $p_0 \in \mathcal{C}, t \notin \overline{\mathcal{D}}$ and $t \to p_0$ it follows

$$k(t, q) = \ln|t - q| \to \ln|p_0 - q| = k(p_0, q).$$

But our aim is to show that $k(t, q) \to k(p_0, q)$ for $t \to p_0$ with t an interior point in $\mathcal{D}$. With this purpose, we will show that, provided p be near enough from p_0, it is possible to determine $t = t(p) \notin \mathcal{D}$ such that

$$k(p, q) - k(t, q) \to 0 \text{ if } p \to p_0. \tag{5.15}$$

Given $p \in \mathcal{D}$, take the line segment that joins p to the point m of $\mathcal{C}$ and which is the one closest to p. Extend such segment up to a point t for which $|t - m| = |m - p|$

holds. For p near enough to p_0, $t \notin \overline{\mathcal{D}}$ and, therefore,

$$k(p, q) - k(t, q) = L_q(g_p - g_t).$$

Remark that

$$g_p(\xi) - g_t(\xi) = \ln \left| \frac{p - \xi}{\xi - t} \right|$$

and, by making use of the hypotheses about $\mathcal{C}$, it may be shown (cf. [21] or the original work, [45]), that when $p \to p_0$ then $t(p) \to p_0$ and $|p - \xi|/|\xi - t| \to 1$, uniformly with respect to $\xi \in \mathcal{C}$. From this it follows that

$$\|g_p - g_t\|_\infty \to 0$$

and, by the continuity of L_q, (5.15) may be obtained.

This way we have succeeded in proving:

The **Dirichlet problem** (5.4) admits a unique solution for arbitrary function f, provided it is a continuous function and the hypotheses on the region $\mathcal{D}$ stated on the opening of this section all hold.

5.3 Open Mapping: Closed Graph

Our road takes us now to some properties of linear mappings:

$$T : B_1 \to B_2,$$

where both $B_1 and B_2$ are Banach spaces.

Consider the **product space**

$$B_1 \times B_2 := \{z := (v_1, v_2);\ v_\iota \in B_\iota, \iota = 1, 2\}$$

equipped with the norm

$$\|z\| = \|(v_1, v_2)\| := \|v_1\| + \|v_2\| \tag{5.16}$$

and the vector space operations **component-wise** defined.

Exercise 5.4 Prove that $B_1 \times B_2$ is a Banach space.

Being T linear, the **graph** of T,

$$G(T) := \{(v_1, v_2) \in B_1 \times B_2;\ v_2 = Tv_1, v_1 \in B_1\},$$

is a subspace of $B_1 \times B_2$. Assuming T continuous results $G(T)$ to be closed. In fact, the conclusion that $G(T)$ is closed has no relation with being $B_1 and B_2$ complete; it is even free from the linearity assumption: it suffices to have T continuous and $B_1 and B_2$ metric spaces.

The converse of this result is precisely the

Closed Graph Theorem (Banach) *Let $T : B_1 \to B_2$ be a linear mapping. Whenever the graph of T is closed on $B_1 \times B_2$ under the norm defined by (5.16), then T is continuous.*

Observe that the function

$$\begin{aligned} f : \mathbb{R} &\to \mathbb{R} \\ x &\to \begin{bmatrix} 0 & x = 0 \\ 1/x & x \neq 0 \end{bmatrix} \end{aligned}$$

has its graph closed, without being continuous. This tells us that linearity of T is essential. Besides, if B_1 is not complete, the conclusion may be false, as indicated by Example 5.2, Sect. 2.11. In the same way, requiring B_2 to be complete may not be disposed of.

Exercise 5.5 All norms

$$\|z\|_p := (\|v_1\|^p + \|v_2\|^p)^{1/p}, 1 \leq p < \infty, \quad \|z\|_\infty := \max\{\|v_1\|, \quad \|v_2\|\}$$

taken on $B_1 \times B_2$ turn out to be equivalent.

A mapping between two metric spaces, $f : M_1 \to M_2$, is continuous if and only if the inverse image $f^{-1}(A)$ of any open set $A \subset M_2$ would be an open set in M_1. Whenever is f 1-1, denoting its inverse by $g : f(M_1) \to M_1$, it is seen that the image of every open set in the domain of g – that is, in the image of f – is an open set in M_1. Thus, g maps the open sets in its domain into open sets. A transformation with such a property is said to be an **open mapping**.

Within the framework of normed spaces, we can count with what states

Exercise 5.6 Given a linear function $f : N_1 \to N_2$, the condition to be f open is to exist $\delta > 0$, $\rho > 0$ such that

$$f(B(0; \delta)) \supset B(0; \rho).$$

Otherwise said, in order to be sure that a given linear transformation is an open mapping, it is enough to assume that the image by f of some open ball contains another open ball.

Recall that every analytic complex variable function, with a connected domain, is necessarily open; otherwise it is a constant, cf. [62, pp. 214]. For $T \in \mathcal{L}(\mathbb{R}, \mathbb{R})$, it also holds that either is it open or a constant, (better said, null). And, given $T \in \mathcal{L}(\mathbb{R}^2, \mathbb{R}^2)$, we have an alternative to generalize what occurs on the real line – either is T open or it is singular – which also holds for every $\mathbb{R}^n$, with $n \geq 2$. As a matter of fact, that is precisely the formulation that holds in general, as long as we take into account that, for a linear operator on a finite dimension space$X$, the hypotheses of being **non-singular** ($ker(T) = \{0\}$) and being an **onto** mapping ($Im(T) = X$) are equivalent:

Open Mapping Theorem *Let T be a mapping from B_1 into B_2, assumed linear, continuous, and* ***onto****. Then T must be an open mapping.*

It is straightforward deducing as one of the consequences of this result the

Isomorphism Theorem (Banach) *Let T be a linear transformation, continuous and* ***1-1*** *from B_1* ***onto*** *B_2. Necessarily T^{-1} ought to be continuous.*

Proof Denote $S := T^{-1} : B_2 \to B_1$. The inverse image of an open set $A_1 \subset B_1$, $S^{-1}(A_1) = T(A_1)$ is then open, as a consequence of the previous theorem as well as the hypotheses made about T. From this, the continuity of T^{-1}follows.

It is shown now that the previous result may as well be proven with the help of the Closed Graph Theorem.

Being closed the graph of the mapping $T : B_1 \to B_2$, it follows that the graph of T^{-1} is closed as well. Indeed, let $(v_n, T^{-1}v_n)$ be a sequence in $G(T^{-1})$ which converges to $(v_0, w_0) \in B_2 \times B_1$. By (5.16), $v_n \to v_0$ and $T^{-1}v_n \to w_0$. From the continuity of T, it may be concluded that $v_n \to Tw_0$ and therefore $v_0 = Tw_0$ or $w_0 = T^{-1}v_0$, which means $(v_0, w_0) \in G(T^{-1})$; in other words, $G(T^{-1})$ is closed.

Conclusion: T^{-1} is continuous.

It deserves to underline that the three theorems stated in the current section are all equivalent. We indicate then how Banach Isomorphism Theorem implies both the Closed Graph and the Open Mapping Theorems.

Exercise 5.7 Verify: if the hypotheses for the Closed Graph Theorem hold for the operator T, by applying to

$$\begin{aligned} \pi_T :\ & G(T) \to B_1 \\ & (x, Tx) \to \pi_T x := x \end{aligned}$$

the Isomorphism Theorem, the continuity of T is deduced, since

$$\|\pi_T x\| = \|x\| + \|Tx\|.$$

Exercise 5.8 Assume for T the conditions on the Open Mapping Theorem. Then $F := ker(T)$ is closed, and, as a consequence, $\tilde{B}_1 := B_1/\sim_F$ is a Banach space. The mapping

$$\left.\begin{aligned} \tilde{T} :\quad & \tilde{B}_1 \to B_2 \\ x^* \in\ & \tilde{B}_1 \to \tilde{T}x^* := Ty,\ \forall y \in x^* \end{aligned}\right]$$

is linear, continuous, 1-1 and onto B_2, thus open, by the Isomorphism Theorem. From this we conclude to be T open.

Let us get back to Theorem 5.1 from Sect. 2.13: it is a straightforward consequence of the Isomorphism Theorem – just take as T the identity.

A more involved consequence of the Closed Graph Theorem is presented in Sect. 5.5. The example that follows shows another application, where we get back to Exercise 2.35 from Sect. 2.12.

Example 5.8 Fourier transform

$$\mathcal{F} : L^1(\mathbb{R}) \to C_a(\mathbb{R})$$

fails to be onto.

It has been verified, on the quoted Exercise 2.35, to be $\mathcal{F}^{-1}$ discontinuous, which implies not being $\mathcal{F}$ onto $C_a(\mathbb{R})$, as this would lead to a contradiction with the Isomorphism Theorem.

As previously announced, this exposition fails to include the proof for the Open Mapping Theorem. We mention, though, that its proof main ingredient is **Baire Category Theorem** on complete metric spaces, cf. [67, pp. 74], or [55, pp. 65]. It is also a consequence of Baire Theorem having we avoided to seek, for Example 2.9, Sect. 2.14, a sequence of continuous functions f_N to approximate $\Psi_{\mathbb{Q}}$. This theorem implies that a function which is the point-wise limit of continuous functions may not present as its set of discontinuity points a collection "as large" as the function $\Psi_{\mathbb{Q}}$ does, namely, the whole real line, cf. [39].

5.4 The Weak Convergence

There exist some information data on elements or subsets from a given normed space N that can be deduced from data about the action of the functionals $\ell \in N^*$ on these elements or subsets. This is what occurs, *e.g.*, with expression (5.3), Theorem 4.1 and Exercise 4.2.

Assume $(x_\jmath)$ to be a convergent sequence on N with limit x. Then for each $\ell \in N^*$ the following holds

$$\ell x_\jmath \to \ell x. \tag{5.17}$$

Conversely, suppose that, for the sequence $(x_\jmath)$ and the vector x, (5.17) holds, for every $\ell \in N^*$. Think about the following question: does **strong convergence**

$$x_\jmath \to x,$$

hold, *i.e.*, is $\|x_\jmath - x\| \to 0$ true?

Example 5.9 Let $e_J := (\delta_J^N)_N \in \ell^2$. By Riesz Representation Theorem, for any $\ell \in (\ell^2)^*$, there exists $v = (v_N) \in \ell^2$ such that

$$\ell e^N = \sum_{N=1}^{\infty} v_J \delta_J^N = v_J.$$

Thus, (e_J) fulfills (5.17) with $x := 0$, since

$$(v_N) \in \ell_2 \Longrightarrow \lim_N v_N = 0.$$

On the other hand, $\|e_J\|_2 = 1$ and on account of this identity, it is impossible for the limit $e_J \to 0$ to hold.

On a certain number of problems, the definition given below is quite important. It is possible to list, among them, the variational formulation for problems from Physics.

It is said that a sequence (x_J) in N **converges weakly** to a vector $x \in N$ if

$$\lim_J \ell x_J = \ell x \text{ for every functional } \ell \in N^* \tag{5.18}$$

holds. Hereby the first point to worry about is the **uniqueness** of this so-called **weak limit**. Suppose that (x_J) is such that its weak convergence simultaneously hold for x and y. The uniqueness of the limit for a sequence of real numbers lets to conclude that, for arbitrary $\ell \in N^*$, the equality below holds

$$\ell x = \ell y.$$

A consequence of Example 5.2, Sect. 4.2 is that $x = y$ then.

Exercise 5.9 Verify: if (x_J) and (y_J) are wekly convergent sequences with limits x e y, respectively, and if the real sequences (α_J) and (β_J) converge respectively to α and β, then we have

$$\lim_J(\alpha_J x_J + \beta_J y_J) = \alpha x + \beta y \text{ (weakly)}.$$

5.5 The Uniform Boundedness Theorem

Every weakly convergent sequence $\{x_n; n \in \mathbb{N}\}$ is **weakly bounded**. In other words, for each arbitrary $\ell \in N^*$, the set of numbers $\{\ell x_n; n \in \mathbb{N}\}$ is always bounded. We have in mind to reach a stronger boundedness relation, namely, one which assures that $\{\|x_n\|; n \in \mathbb{N}\}$ is bounded, as occurs with the strongly convergent sequences. By taking (5.3) into account, it amounts to estimate

$$\{\ell x_n; n \in \mathbb{N}, \ell \in N^*, \|\ell\| = 1\},$$

or, making use of the notation in Sect. 3.6,

$$\{J_{x_n}\ell; n \in \mathbb{N}, \ell \in N^*, \|\ell\| = 1\}.$$

For each fixed n, the relations

$$|J_{x_n}\ell| \leq \|J_{x_n}\|_{N^{**}}\|\ell\|_{N^*} = \|x_n\|_N\|\ell\|_{N^*} = \|x_n\|_N$$

hold and, by the same token, for any fixed ℓ

$$|J_{x_n}\ell| = |\ell x_n| \leq C_1(\ell),$$

since (ℓx_n) converges.

Observe that the constants $C_1(\ell)$ and $C_2(n) := \|x_n\|$ are independent of n as well as of ℓ, respectively. Our aim is reaching a unique constant C such that

$$C_1(\ell) \leq C \ \forall \ell \in N^*, \|\ell\| = 1,$$
$$C_2(n) \leq C \ \forall n \in \mathbb{N}.$$

Let us have a look on a more general framework.

Uniform Boundedness Theorem (Banach–Steinhaus) *Let $(T_\alpha)_{\alpha\in\Delta}$ be a family of linear continuous operators, $T_\alpha : B \to N$, with Δ as an arbitrary set, B a Banach space and N a normed space. Denoting by $S_1 := \{x \in B; \|x\| = 1\}$, suppose to be fulfilled the following point-wise bounds:*

$$\sup_{\alpha\in\Delta} \|T_\alpha x\| \leq C_1(x), \text{ for each } x \in S_1, \tag{5.19}$$

$$\sup_{x\in S_1} \|T_\alpha x\| \leq C_2(\alpha), \text{ for each } \alpha \in \Delta. \tag{5.20}$$

It is then guaranteed the existence of a constant C related to a uniform bound:

$$\sup_{\{x\in S_1 \alpha\in\Delta\}} \|T_\alpha x\| \leq C. \tag{5.21}$$

The presented formulation emphasizes some kind of symmetry between the variables x and α. It is more usual to express this result in the form

> Every pointwise bounded set $\{T_{\alpha\in\Delta}\} \subset \mathcal{L}(B, N)$ is equicontinuous, as long as it is uniformly bounded.

Proof Let $L^{\infty}(\Delta; \tilde{N})$ be the Banach space of the bounded functions from Δ on $\tilde{N}$, where $\tilde{N}$ denotes the completion of N, equipped with the norm

$$\|f\|_{L^{\infty}(\Delta;\tilde{N})} := \sup_{\alpha\in\Delta} \|f(\alpha)\|_{\tilde{N}}.$$

Consider the linear mapping

$$\left.\begin{array}{l} S : B \to L^{\infty}(\Delta; \tilde{N}) \\ \quad x \to Sx := h_x \end{array} \text{where } \begin{array}{l} h_x : \Delta \to \tilde{N} \\ \quad \alpha \to h_x(\alpha) := T_\alpha x \end{array}\right].$$

Observe that if S is bounded, as long as

$$\begin{aligned} \|S\| := \sup_{\|x\|_B=1} \|Sx\|_{L^{\infty}(\Delta;\tilde{N})} &= \sup_{\|x\|_B=1} \sup_{\alpha\in\Delta} \|h_x(\alpha)\|_{\tilde{N}} \\ &= \sup_{\|x\|_B=1} \sup_{\alpha\in\Delta} \|T_\alpha x\|_{\tilde{N}}, \end{aligned}$$

we will have obtained (5.21) with $C := \|S\|$. In order to reach the proof of the continuity of S, we demand help from the Closed Graph Theorem.

Let (x_n, Sx_n) be a sequence of points on the graph of S, and suppose its convergence to (x, y), with $y \in L^{\infty}(\Delta; \tilde{N})$. To show the graph of S is closed, we must prove that $y = Sx$.

Since $x_n \to x$, for each $\alpha \in \Delta$ occurs the convergence

$$T_\alpha x_n \to T_\alpha x,$$

which means that

$$\lim_n \|T_\alpha x_n - T_\alpha x\|_N = 0. \tag{5.22}$$

On the other hand,

$$\begin{aligned} 0 = \lim_n \|Sx_n - y\|_{L^{\infty}(\Delta;\tilde{N})} &= \lim_n \sup_{\alpha\in\Delta} \|h_{x_n}(\alpha) - y(\alpha)\|_{\tilde{N}} \\ &= \lim_n \sup_{\alpha\in\Delta} \|T_\alpha x_n - y(\alpha)\|_{\tilde{N}}. \end{aligned} \tag{5.22'}$$

From (5.22) and (5.22'), it results that $T_\alpha x_0 = y_0(\alpha)$, that is, $Sx_0 = y_0$, since

$$[Sx_0](\alpha) = h_{x_0}(\alpha) = T_\alpha x_0 = y_0(\alpha).$$

Concluded the proof, we can now state the following result, discussed before having presented the Banach-Steinhaus Theorem.

Theorem 5.3 *Let (x_n) be a weakly convergent sequence in the normed space N. We then assure that $\{\|x_n\|; n \in \mathbb{N}\}$ is a bounded set on the line, and we say that (x_n) is strongly bounded.*

Observe: no need to assume N to be complete, as the Banach-Steinhaus Theorem is applied in the space N^*, which is always a Banach space.

Exercise 5.10 A sequence (x_j) in N is said to be **weakly Cauchy** if, for arbitrary $\ell \in N^*$, the real sequence (ℓx_j) is a Cauchy sequence. Prove that every weakly Cauchy sequence is strongly bounded.

Exercise 5.11 If the sequence of bounded operators $T_n : B \to N$ **converges pointwise** toward an operator T, which means

$$T_n x \to Tx, \text{ for each } x \in B,$$

then the operator T is linear and continuous.

5.5.1 An Application to Numerical Schemes

In 1928, Richard Courant, Kurt O. Friedrichs, and Hans Lewy, in the article [20], published a benchmark for the rigorous curtain opening of Numerical Analysis for partial differential equations.[4] Therewith they formalized the theory for finite differences algorithms, as long as they made clear the sometimes surprising behavior of some numerical schemes for **pde**'s. They introduced the concept of **stability**, showing how this property is strongly related to **convergence**.

Passed around 30 years, Peter D. Lax observed, cf. [48], that the relation between these two concepts is still deeper – they are equivalent, as expressed by the result exposed in the sequel, which turns out to be an application of Banach-Steinhaus Theorem.

Take the **heat diffusion linear equation** as an example which illustrates the introduced concepts.[5] The initial value problem – IVP – or Cauchy problem,

$$\left.\frac{\partial u(x,t)}{\partial t} - \nu \frac{\partial^2 u(x,t)}{\partial x^2} = 0\right] \begin{array}{c} -\infty < x < \infty \\ 0 \leq t \leq T \end{array} \tag{5.23}$$

$$u(x,0) = \phi(x) \quad -\infty < x < \infty \tag{5.24}$$

may be formulated in the context of function spaces as

---

[4] Also available through a version to English in [27].

[5] We borrow the exposition in [57], but bringing in a lot of simplifying.

$$\begin{aligned} \frac{d}{dt}u(t) &= Lu(t) \\ u(0) &= \phi \end{aligned}, \tag{5.25}$$

for $u \in C^1(0, T; H^1(\mathbb{R}))$ and the differential operator $L : X \to H^1(\mathbb{R})$ defined in X, a subspace of $H^1(\mathbb{R})$ to be duly stated.

Among other points to be considered ahead the choice of this space, let us mention that it ought to include the boundary conditions imposed to u – whenever one deals with the mixed problem. Such a choice must also assure that, with such a formulation, (5.25) states a **well-posed problem**, in the sense of Hadamard. This means, informally, that the collection of allowed solutions is *large enough*, and they depend in a unique and continuous way from the given data.

With a more precise saying, let $\mathcal{X} \subset H^1(\mathbb{R})$ be such that, if $\phi \in \mathcal{X}$, (5.25) admits a solution, being thus defined a family of operators

$$\left.\begin{aligned} \mathcal{E}(t) : \mathcal{X} &\to H^1(\mathbb{R}) \\ \phi &\mapsto \mathcal{E}(t)\phi := u(t) \end{aligned}\right], t \in]0, T]. \tag{5.26}$$

It is said that this is a well-posed problem, in the sense of Hadamard, if it is guaranteed to hold

a. **Existence** $\mathcal{X}$ dense in $H^1(\mathbb{R})$
b. **Uniqueness** $\mathcal{E}(t)$ uniquely defined
c. **Continuity** $\mathcal{E}(t)$ uniformly bounded

The first condition, a stronger formulation than the original one, by Hadamard, assures the existence of **generalized solutions**. If the function $\psi \in H^1(\mathbb{R}) \setminus \mathcal{X}$, it may be approximated by a sequence in $\mathcal{X}$, and, due to *c*., the solutions whose initial values are the elements of this sequence form a sequence which also converges. The limit thus obtained is then assigned to the *solution* which has ψ as its initial value. Again, all amounts to an application of the Principle of the Continuous Extension, this time to the operators $\mathcal{E}(t)$.

The third condition is known as **continuous dependence on initial data**. It assures that small errors – either from the measurements or the calculations – associated to the initial value will lead to small deformations on the system response.

Numerical approximations for the solution u of (5.25) are built with finite difference through the following steps.

First choose the discretization parameter $k := \Delta t := T/N$, then calculate, for $t^n := nk$ and $n = 1, \ldots, N$, simulations for $u^n := u(t^n)$, denoted by U^n and initially defined on a finite set of points $x_j := jh$, $j \in \mathbb{N}$, with $h := \Delta x > 0$, by an evolution scheme.[6]

The approximations U^n are generated then in $H^1(\mathbb{R})$ via some interpolation starting from the values calculated in (x_j, t^n) and denoted U_j^n, taking into account

[6] It is just to simplify the presentation that we restrict to using uniform meshes – the discretization sub-intervals have the same length, just for x as for t.

$1 \le n \le N$, $\jmath \in \mathbb{N}$. We look forward to guarantee $U_\jmath^n \sim u_\jmath^n := u(x_\jmath, t^n)$, which means to be dealing with a **convergent scheme**, that will be shortly made precise.

The considered algorithm follows the time evolution sequentially according to

$$\begin{bmatrix} U_\jmath^0 & := \phi(\jmath h) & \jmath \in \mathbb{N} \\ U^{n+1} & := \mathbf{B}(h,k)U^n & 0 \le n < N \end{bmatrix}, \tag{5.27}$$

being $\mathbf{B}$ the chosen finite differences operator. For example, maybe the simplest scheme for (5.23) is

$$\frac{U_\jmath^{n+1} - U_\jmath^n}{k} = \sigma \frac{U_{\jmath+1}^n - 2U_\jmath^n + U_{\jmath-1}^n}{h^2}, \tag{5.28}$$

from which it follows, by denoting with $\tau_{\pm h}$ the right, resp. left, translation,

$$\mathbf{B}(h,k)U^n := U^n + \frac{k}{h^2}\left(\tau_{+h}U^n - 2U^n + \tau_{-h}U^n\right).$$

To get hold of a numerical scheme which simulates, or discretly models, a given differential equation means that the solutions of the finite differences relation it requires somehow approximate the solution for the given differential equation. Being the scheme **compatible** with the original equation, the solution should also be *almost* a solution to the differences equation – and such is the condition to impose so as to define a **consistent** algorithm. Through other words, as long as it is expected that

$$\frac{U^{n+1} - U^n}{k} \sim LU^n,$$

for I the identity, **consistency is defined** by requiring, for the solutions of (5.25), or at least for a family which is dense on this set, the condition

$$\lim_{h,k\to 0} \left\| \left[\frac{\mathbf{B}(h,k) - I}{k} - L \right] u(t) \right\|_{H^1(\mathbb{R})} = 0, \forall t \in [0, T]. \tag{5.29}$$

From now on it will be supposed that the discretization meshes hold a dependence relation, *v.g.*, $h = g(k)$. The relation

$$U^{n+1} = \mathbf{B}U^n = \ldots = \mathbf{B}^{n+1}U^0$$

implies to be demanded in the calculations powers of the involved operators $\mathbf{B} = \mathbf{B}(k)$, and consequently the numerical solutions built in by the scheme under exam depend on two parameters, k and n. These remarks let us to introduce the

Definition (Convergence) The numerical scheme (5.27) turns out to be convergent if, given any sequences of integers (n_ℓ), (N_ℓ), with the properties

$$\left.\begin{array}{l} a.\ n_\ell, N_\ell \xrightarrow{\ell} \infty \\ b.\ \text{being } k_\ell := T/N_\ell,\ n_\ell k_\ell \xrightarrow{\ell} t \text{ for some } t \in]0, T] \end{array}\right],$$

the limit below always holds

$$\|\mathbf{B}(k_\ell)^{n_\ell}\phi - \mathcal{E}(t)\phi\| \xrightarrow{\ell} 0,\ \phi \in \mathcal{X}. \tag{5.30}$$

A concept that rested unknown until [20] has been published, when it was then formalized, is the **stability** of a numerical scheme. The operators $\mathbf{B}^n$, for n large, are bound to approximate the operator $\mathcal{E}$ for solving the IVP. They ought, therefore, to keep themselves bounded, more than that, **uniformly bounded**. That is exactly what is required by the

Definition (Stability) A scheme (5.27) is considered to be stable if

$$\exists C > 0;\ \|\mathbf{B}(k_\ell)^n\| \leq C,\ \forall n \in \mathbb{N},\ k_\ell \xrightarrow{\ell} 0. \tag{5.31}$$

Another reason to require the stability of the numerical schemes: (5.31) also implies that the distortions introduced in the initial data, among them those brought in by rounding errors, despite being reactivated by successive powers of $\mathbf{B}$, are kept bounded.

Definition (5.29) establishes a local constraint, since it encompasses to estimate the evolution of the approximate solution from level n to level $n+1$, while (5.30) and (5.31) are global. On the other side, opposite to the two other definitions, (5.31) keeps track only of the numerical scheme, not to mention the considered equation. The strong relation that links these three properties is the core of

The Lax Equivalence Theorem. *For a well-posed initial value problem, a consistent finite difference scheme is convergent if and only if it is stable.*

By contradiction, for a convergent scheme, let us suppose to be possible finding $\phi \in \mathcal{X}$ that originates a sequence of non-bounded approximations, thus contradicting (5.31),

$$\|\mathbf{B}(k_\ell)^{n_\ell}\phi\| \to \infty. \tag{5.32}$$

Then Bolzano-Weierstrass theorem implies to be possible finding a subsequence for which

$$n_{\ell'}k_{\ell'} \xrightarrow{\ell'} \bar{t} \text{ for some } \bar{t} \in [0, T].$$

Analogously condition (5.30) implies to be convergent this subsequence, and, therefore, (5.32) cannot hold. On this track, we conclude that, given $\phi \in \mathcal{X}$, there exists a constant $C_1(\phi)$ for which

$$\|\mathbf{B}(k)^n \phi\| \leq C_1(\phi), \forall n \in \mathbb{N}, k_\ell \xrightarrow{\ell} 0.$$

But this is exactly the condition under which the Banach-Steinhaus theorem assures the uniform boundedness (5.31), and, as a consequence, **convergence implies stability**.

The opposite sense of the proof is standard in Numerical Analysis; see [14, 57].

The concepts introduced in (5.29), (5.30), and (5.31) may hold (or fail to) independently of the relation between h and k. It is then said that such a scheme is **unconditionally** – or conditionally – consistent, convergent or stable. For example, the scheme (5.28) is unconditionally consistent, but it is stable – thus convergent – if and only if

$$h/k^2 = \Delta t/(\Delta x)^2 \leq \nu. \tag{5.33}$$

Suppose a scheme to be unconditionally stable but conditionally consistent. Such a coupling, which seems to be helpful, may lead to computational risks, as illustrated by the reasoning that follows.

Consider an arbitrary sequence of numerical approximations. The scheme stability implies this sequence boundedness and, by Theorem 6.2, the weak convergence of some of its subsequences. This convergence may lead to evaluate that the discretization parameters keep the required relation for consistence of the scheme being used. It occurs, though, that in some problems, we fail to deal with a quite "clear" condition like the one described in (5.33). It may then occur that the reached convergence – and in practical terms, tested in some spot of the computer program – may be as well generating approximations to another problem; see [24] for details.

Chapter 6
Compactness

6.1 Introduction

Given $X \subset M$ (= metric space), it is said that X is **compact** if, from each sequence $\{x_n\}$ of elements from X, a subsequence $\{x_{n_k}\}$ can be extracted which converges to some $x_0 \in X$.

Exercise 6.1 Let $T : M_1 \to M_2$ be a continuous mapping, being $M_1 and M_2$ metric spaces. Whenever is $K \subset M_1$ compact, $T(K) \subset M_2$ is also necessarily compact.

The ball $B[0, \pi^{1/2}] \subset L^2(0, 2\pi)$ is not compact, as stated on Exercise 2.36. On the other hand, in every finite dimensional normed space, a closed and bounded set is always compact, and, in fact, the following characterization discovered by F. Riesz, cf. [72, pp. 85], is valid:

Theorem 6.1 *On a normed space N, the closed balls are compact in the sense of the norm – or **strongly compact** – if and only if N is finite dimensional.*

Bolzano-Weierstrass theorem claims that the compact sets in $\mathbb{R}^p$ are the bounded and closed sets. Given a normed space N with finite dimension p, it is **homeomorphic** to $\mathbb{R}^p$, which means, by definition, that there *exists a linear and continuous mapping* $T : N \to \mathbb{R}^p$, *1-1 and onto, such that* T^{-1} *is also continuous.*[1] Consequently, the bounded and closed subsets of N, since they exhibit a homeomorphic correspondence with the bounded and closed sets from $\mathbb{R}^p$, are compact as well.

We suggest reading [72] for what still remains to be proved.

[1] Any linear mapping from N onto $\mathbb{R}^p$, being 1-1, is necessarily a homeomorphism, according to the remark that follows Exercise 2.30.

C. A. de Moura, *Functional Analysis Tools for Practical Use in Sciences and Engineering*, https://doi.org/10.1007/978-3-031-10598-2_6

6.2 Compactness in C^0 and L^p

It is a straight deduction from the definition that every compact set is closed and bounded. From Theorem 6.1 above, in the infinite dimensional spaces, we must get hand of other conditions so as to assure compactness for any of its subsets. So far everything points out that a general characterization lacks to be available. What we happen to recognize is, for some spaces, particular conditions to enroll their compact sets.

To open the display, consider the space $N := C^0[0, 1]$ with the norm $\|\cdot\|_\infty$. Assume $F \subset N$ to be one of its compact subsets. Given $\{f_n\}$, an arbitrary sequence in F, there exists a subsequence $\{f_{n_k}\}$ such that

$$\lim_{k\to\infty} \|f_{n_k} - f\|_\infty = 0,$$

for a particular $f \in$F. Since f is uniformly continuous, for any $\epsilon > 0$, there exists $\delta = \delta(\epsilon, f) > 0$ for which

$$x_1, x_2 \in [0, 1], |x_1 - x_2| < \delta \Longrightarrow |f(x_1) - f(x_2)| < \epsilon. \tag{6.1}$$

On the other hand, it is possible to determine $K = K(\epsilon)$ such that, if $k \geq K$, it is verified that

$$|f_{n_k}(x) - f(x)| < \epsilon, \forall x \in [0, 1].$$

We thus conclude: if $|x_1 - x_2| < \delta$ and $k \geq K$,

$$\begin{aligned} |f_{n_k}(x_1) - f_{n_k}(x_2)| \leq |f_{n_k}(x_1) - f(x_1)| + \\ |f(x_1) - f(x_2)| + \quad |f(x_2) - f_{n_k}(x_2)| < 3\epsilon. \end{aligned}$$

Therefore, the functions f_{n_k} are not only uniformly continuous, but they happen to be what is called **equicontinuous**,[2] which means for any $\epsilon > 0$, there exists $\delta = \delta(\epsilon) > 0$ (which depends only on ϵ), for which

$$|x_1 - x_2| < \delta \Longrightarrow |f_{n_k}(x_1) - f_{n_k}(x_2)| < \epsilon, \forall k \in \mathbb{N}. \tag{6.2}$$

It can be deduced that such a property holds not only for the subsequence $\{f_{n_k}\}$ but for the whole set[3] F, and this way we get hold of the following characterization of the compact sets in $C^0[0, 1]$.

Theorem (Arzelà–Ascoli) *A bounded and closed subset F of $C^0[0, 1]$ is compact if and only if it is equicontinuous.*

[2] Or, more properly, an **equicontinuous set**.

[3] This means that in (6.2) one can exchange $\forall k \in \mathbb{N}$ with $\forall f \in F$.

This result stays valid when $C^0[0, 1]$ is replaced by $C^0(K)$, with K an arbitrary compact, not necessarily contained in $\mathbb{R}$, but in any metric space.

Proof We reason by contradiction, assuming that for the compact F does not hold the equicontinuity property. It would then be possible to determine a number $\bar{\epsilon}$ for which no corresponding $\delta > 0$ would satisfy (6.1) for every $f \in F$. Saying with other terms, for each $\delta > 0$ being chosen, it would be possible to find a function within F for which (6.1) would fail to hold. Now, this fact would imply the existence of a sequence $\{f_n\}$ in F built from the values given by $\delta = \delta_n := 1/n$. Such a sequence would not admit any equicontinuous subsequence. And this would then imply that no convergent subsequence could be extracted from itself.

In order to prove that a bounded, closed, and equicontinuous set in $C^0[0, 1]$ must be compact requires more creative steps. It makes use of the so-called Cantor diagonal process.

Choose a countable dense set in [0,1] – for example, the rational numbers on that interval. Denote this set by $\{q_1, q_2, \ldots\}$, and let $\{f_j\}$ be an arbitrary sequence in F. Being F bounded in the norm $\|\cdot\|_\infty$, the real sequence $\{f_j(q_1)\}$ is bounded, so that by Bolzano-Weierstrass theorem, it is possible to determine a subsequence $\{f_{1_j}\}$ from $\{f_j\}$ such that $\{f_{1_j}(q_1)\}$ is convergent. By the same reasoning, $\{f_{1_j}(q_2)\}$ is bounded, and from this we get $\{f_{2_j}\}$, subsequence of $\{f_{1_j}\}$ with $\{f_{2_j}(q_2)\}$ showing convergence.

Follow then the same procedure for each one of the points q_n, and define $g_n := f_{nn}$. Fixed m, $\{g_j\}_{j\geq m}$ is then a subsequence of $\{f_n\}$, and therefore, $\{g_j(q_m)\}$ is convergent.

On the next step it is proven that $\{g_j\}$ **converges uniformly**. Since F is equicontinuous, choose an arbitrary value for $\epsilon > 0$ to determine $\delta(\epsilon) > 0$ such that (6.1) holds for any $f \in F$. As long as k is large enough and $\mathcal{Q}_k := \{q_1, \ldots, q_k\}$, for any $x \in [0, 1]$, we verify that $\text{dist}(x, \mathcal{Q}_k) < \delta$. Let then $x \in [0, 1]$ be fixed and $q = q(x) \in \mathcal{Q}_k$ satisfying $|x - q| < \delta$. We will have then:

$$
\begin{aligned}
|g_m(x) - g_m(x)| \leq |g_m(x) - g_m(q)| + \\
|g_m(q) - g_n(q)| + |g_n(q) - g_n(x)|.
\end{aligned} \tag{6.3}
$$

Since the set $\mathcal{Q}_k$ is finite, it is possible to determine $N = N(\epsilon)$ such that

$$
m, n \geq N \Longrightarrow |g_m(q) - g_n(q)| < \epsilon, \ \forall g \in \mathcal{Q}_k.
$$

From the equicontinuity of F, the first and third terms in the right-hand side of (6.3) have both ϵ as an upper bound. The conclusion tells that $\{g_n\}$ is a **uniformly Cauchy** sequence, *i.e.*, it is a Cauchy sequence with respect to the norm $\|\bullet\|_\infty$. But $C^0[0, 1]$ is complete for this norm, and F is closed, so these facts allow to deduce convergence of $\{g_n\}$ to some $g \in F$. The proof ended.

Example 6.1 Take $K \subset C^0[0,1]$ bounded. Suppose that every $f \in K$ is differentiable and that there exists a constant M for which the inequalities $\|f'\|_\infty \leq M, \forall f \in K$ all hold. The compactness of $\bar{K}$ follows then through the Mean Value and the Arzelà-Ascoli theorems.

Example 6.2 By using the consequences of Exercise 2.28, it may be deduced to be bounded and equicontinuous the closed unity ball from $H^1(0,1)$ in $(C^0[0,1], \|\bullet\|_\infty)$; thus it is a compact set.

Exercise 6.2 Let $X := \{f : \mathbb{R} \to \mathbb{R}, f \text{ bounded and continuous}\}$, with the norm $\|\cdot\|_\infty$, and let

$$g(x) := \begin{bmatrix} 0 & x \leq 0 \\ x & 0 \leq x \leq 1 \\ 1 & 1 \leq x \end{bmatrix}.$$

Verify that

$$F := \{f; f(x) = g(x-n), n \in \mathbb{N}\}$$

is closed, bounded, and equicontinuous in X but lacks to be compact.

The exercise above intends to illustrate that, in order to assure the compactness for a set F of functions with an unbounded domain, it is mandatory also to impose conditions upon the asymptotic behavior of the elements of F, for $|x| \sim \infty$. This is shown on condition (ii) from the following:

Theorem (Fréchet-Kolmogorov) *Fixed $p \in [1, \infty)$, any closed and bounded subset F of $L^p(\mathbb{R})$ is compact if and only if:*

$$\left.\begin{matrix} (i)\ \lim_{t\to 0} \int_{\mathbb{R}} |f(s+t) - f(s)|^p ds = 0 \\ (ii)\ \lim_{\alpha\to\infty} \int_{|s|>\alpha} |f(s)|^p ds = 0 \end{matrix}\right] \begin{matrix} \textit{uniformly} \\ \textit{for } f \in F \end{matrix}.$$

Condition (i) may be thought of as an "equicontinuity measure" for the generalized functions from F. You may visit the proof on [72, pp. 275].

6.3 The Weak* Convergence

The notions of weak and strong convergence have been previously exposed. On the dual spaces a third concept is employed, the weak* convergence.

Let N be a normed space and N^* its dual. It is said that a sequence of functionals (ℓ_j) in N^* presents a **weak* convergence** to $\ell \in N^*$ if, for every vector $x \in N$, $\ell_j x \to \ell x$.

It turns out this definition to be equivalent to require, for any $F \in J(N)$, that $F\ell_j \to F\ell_0$, where J is the mapping which allows to identify N to the bi-dual N^{**}. This way, whenever is N reflexive, weak* convergence is the same as weak convergence.

As long as the weak* limit of a sequence (ℓ_j) exists, it is unique. On the contrary, suppose that, simultaneously,

$$\ell_j \overset{*}{\to} \ell_0 \text{ and } \ell_j \overset{*}{\to} \ell_0',$$

this notation standing for the weak* convergence. By definition, $\ell_0 x = \ell_0' x$, for every $x \in N$, thus $\ell_0 = \ell_0'$. All standard properties of limit processes behavior may be straightly verified.

Observe that on $\mathcal{D}'(0, T)$ and $\mathcal{D}'(0, T; B)$, it was introduced exactly the weak* convergence. Both theorems stated in the sequel justify to deal with the concepts of weak and weak* convergence.

Theorem 6.2 *If B is a reflexive Banach space and (x_j) a bounded sequence, the latter contains a subsequence (x_{j_k}) which is weakly convergent.*

Therefore the bounded and closed sets in B are **weakly compact**.

Theorem 6.3 *Let N be an arbitrary normed space. Then its bounded and closed subsets are weakly* compact. In another saying, given $S \subset N^*$, bounded and closed, and chosen any sequence (ℓ_j) in S, it may be determined a subsequence (ℓ_{j_k}) and $\ell \in S$ such that $\ell_{j_k} \overset{*}{\to} \ell$.*

The main appeal of Theorem 6.2 rests in applications to the spaces $H^k(\Omega)$ and $L^p(\Omega)$, $1 < p < \infty$. As regards to Theorem 6.3, it is usually applied to $L^\infty(\Omega) = [L^1(\Omega)]^*$. We will restrict ourselves to the

Proof for Theorem 6.2 We will suppose that B is separable. This will simplify the proof and has a reasonable justification: a huge chunk of the function spaces we work with hold such a property.

By Theorem 6.3 on Sect. 6.2, B^* is separable, since its dual $B^{**} = J(B)$ is so. Let (ℓ_j) be dense in B^* and let (x_j) be a bounded sequence in B. As a consequence $(\ell_1 x_j)$ is bounded, and, therefore, there exists a subsequence (x_j^1) such that $(\ell_1 x_j^1)$ is convergent. Now, $(\ell_2 x_j^1)$ is bounded as well, and, thus, it is possible to extract a subsequence (x_j^2) from (x_j^1) such that $(\ell_2 x_j^2)$ converges.

On this way, for each integer $p > 1$, we obtain (x_j^n), a subsequence of (x_j^{p-1}), such that $(\ell_k x_j^p)$ is convergent, for $1 \le k \le p$.

Thus, the diagonal sequence $(x_j') := (x_j)$ satisfies the condition of being $(\ell_k x_j')$ convergent, for $k = 1, 2, \ldots$.

Let now be $\ell \in B^*$ arbitrary and $\epsilon > 0$ be given. The inequality

$$|\ell x_n' - \ell x_m'| \le \|\ell - \ell_k\| \|x_n'\| + |\ell_k x_n' - \ell_k x_m'| + \|\ell_k - \ell\| \|x_m'\|$$

implies to be $(\ell x'_j)$ a Cauchy sequence, thus convergent. Consequently, $(\ell x'_j)$ is convergent, no matter which $\ell \in B^*$is chosen, and now all is left to show is that there exists $x_0 \in B$ for which

$$\ell x_0 = \lim_j \ell x'_j.$$

Consider $F_j := Jx'_j \in B^{**}$. The functional

$$F_0\ell := \lim_j F_j\ell = \lim_j \ell x$$

is linear and continuous, by Banach-Steinhaus Theorem. This way we have reached the conclusion that $F_0 \in B^{**} = JB$, and thus $F_0 = Jx_0$, for some $x_0 \in B$. It amounts to a short job to verify that

$$\lim_j x'_j = x_0 (\text{weak sense}).$$

The converse to Theorem 6.2 is the

Eberlein-Shmulyan Theorem: *Suppose that every bounded sequence in the Banach space B owns a weakly convergent subsequence. For sure B is then reflexive.*

The long and tedious proof for this result may be found in [72, pp. 141].

6.4 Rellich and Immersion Theorems

A **linear operator** $T : N_1 \to N_2$ is said to be **compact** if its image for any bounded set (from N_1) has a compact closure (in N_2). Alternatively if, from each bounded sequence (x_j) in N_1, it is possible to obtain a subsequence (x'_j) such that (Tx'_j) is convergent in N_2.

Example 6.3 Consider the Example 6.3 from Sect. 2.12. Arzelà-Ascoli Theorem assures compactness for the operator:

$$f \to \int_0^x f(s)ds.$$

Example 6.4 Take now the operator

$$\left.\begin{array}{ll} \imath : H^1(0,1) \to C^0(0,1) \\ f \quad\to \imath(f) := f \end{array}\right].$$

With (2.14) and again Arzelà-Ascoli Theorem, we deduce the compactness of $\imath$.

This fact is described by saying that $H^1(0, 1)$ is **compactly immersed** in $C^0(0, 1)$. It exhibits an example of a series of results known as **immersion theorems for the Sobolev spaces**. Essentially, they pass along the information that the functions on the spaces $H^k(\Omega), \Omega \subset \mathbb{R}^n$, are much more regular than their definitions would make us to expect.

Theorem (Relich) *Let $\Omega \subset \mathbb{R}^n$ be open and bounded. Given any sequence $(f_\jmath)$ in $H_0^k(\Omega)$, $k \geq 1$, there exists a subsequence $(f_{\jmath_p})$ which converges in $H_0^{k-\ell}(\Omega)$, $1 \leq \ell \leq k$. Rephrasing: compactness holds for*

$$\begin{array}{ll} \imath : H_0^k(\Omega) \to H_0^s(\Omega) \\ \quad f \quad \to \imath(f) := f \end{array}, \text{ being } 0 \leq s \leq k-1.$$

Example 6.5 The condition which requires to be Ω bounded cannot be forgotten. Indeed, let $f \in H^k(\mathbb{R}^n)$ and $x_0 \in \mathbb{R}^n$ be an non-null vector. In such a case,

$$f_\jmath(x) := f(x - \jmath x_0), \jmath \in \mathbb{N}$$

is a bounded sequence which, in $H^s(\mathbb{R}^n)$, fails to admit a converging subsequence, no matter the value of s.

For unbounded regions we may use the

Theorem 6.4 *Take $\Omega \subset \mathbb{R}^n$ an open set and $(f_\jmath)$ a bounded sequence in $H^1(\Omega)$. In this case it is possible to choose a subsequence $(f_{\jmath_p})$ and a function f_0 in $L^2_{loc}(\Omega)$ for which, in each compact $K \subset \Omega$,*

$$\int_K |f_{\jmath_p}(x) - f_0(x)|^2 dx \overset{p}{\to} 0.$$

Rellich theorem remains valid for $H^k(\Omega)$ (replacing $H_0^k(\Omega)$) provided $\partial\Omega$ own some regularity, see [32, pp. 31].

With regard to the smoothness level shown by the elements of $H^k(\Omega)$, a valid tool is the

Theorem 6.5 *Let $k > n/2$ be an integer. The elements from $H^k(\mathbb{R}^n)$ are then continuous functions. To be more precise, for each $f \in H^k(\mathbb{R}^n)$, there exists $g \in C^0(\mathbb{R}^n)$ such that $g = f$ ae.*

With still a greater degree of generality, if $k - \ell > n/2$, except for changes on a set of null measure, we have $f \in C^\ell(\mathbb{R}^n)$ and

$$\|f\|_{\ell,\infty} \leq C\|f\|_{k,2},$$

for some constant $C = C(\ell, k)$, which does not depend on f.

A corollary may be gotten from Theorem 6.3: we may reach to the same conclusions for $H_0^k(\Omega)$, being Ω an open subset from $\mathbb{R}^n$. Again, analogous properties for $H^k(\Omega)$ may be proved, as long as $\partial\Omega$ own enough smoothness, cf. [32, pp. 30], or [54, pp. 80].

Chapter 7
Function Derivatives in Normed Spaces

7.1 Introduction

Many nonlinear problems are sometimes treated as **a perturbation** for linear ones. Such an approach allows a simpler track, as we deal with a more familiar structure. As long as we know the latter with more details, we can migrate conclusions to those tougher ones. That is the idea which lies, for example, when we approximate a real function by its Taylor expansion restricted to the first-order term, namely by employing the Mean Value Theorem:

$$f(x_0 + h) = f(x_0) + f'(x_0 + \theta h)h, \ \theta \in (0, 1)$$

in the approximated form

$$f(x_0 + h) \approx f(x_0) + f'(x_0)h.$$

The knowledge of the derivatives of a function lets us to replace, locally, such a function by a linear approximation. This section brings the concept of derivative from a more general viewpoint, so as to encompass vector functions.

Let M and N be normed spaces, and let $f : D \subset M \to N$ be an arbitrary function.[1] It is said that f is differentiable in Fréchet sense at the point x_0 of its domain, whenever there exists a bounded operator $T = T(x_0, f) : M \to N$ such that

$$[f(x_0 + h) - f(x_0) - Th] = \boldsymbol{o}(\|h\|), \ h \to 0. \tag{7.1}$$

[1] In fact, we shall use always this notation, but in the sequel the domain of the functions may be restricted to an open portion of M, or of the corresponding space.

C. A. de Moura, *Functional Analysis Tools for Practical Use in Sciences and Engineering*, https://doi.org/10.1007/978-3-031-10598-2_7

In (7.1) we employ the notation from (4.59), meaning that

$$\lim_{h\to 0} \frac{\|f(x_0+h)-f(x_0)-Th\|_N}{\|h\|_M} = 0.$$

It may be quickly verified that

(*a*) There exists at most a bounded operator which fulfills (7.1). Therefore, it is consistent to mention T as **the** (Fréchet) derivative for f at the point x_0, which is denoted then by $f'(x_0)$.
(*b*) Being f a constant, its derivative is the null operator:

$$f'(x_0) = 0, \forall x_0 \in M.$$

(*c*) When f is a linear operator, then $f'(x_0)$ exists for any $x_0 \in M$ and, in fact,

$$f'(x_0) = f, \forall x_0 \in M.$$

(*d*) If f is differentiable at the point x_0, it is continuous at this point.
(*e*) Being f differentiable at x_0, for any $z \in M$ it holds:

$$\lim_{\lambda\to 0} \frac{f(x_0+\lambda z)-f(x_0)}{\lambda} = f'(x_0)z. \tag{7.2}$$

Independently of being f differentiable at x_0 or not, as long as the limit in (7.2) exists, it is named the **directional derivative** – or Gâteaux derivative – for f at x_0 on the direction of z, denoted $(\partial f/\partial z)(x_0)$. It is clear that if $y := \alpha z$, then $(\partial f/\partial y)(x_0) = (\partial f/\partial z)(x_0)$.

We present the proofs for (a) and for (b).

Suppose that for T_1 and T_2, (7.1) holds, it follows that, given $\epsilon > 0$ there exists $\delta = \delta(\epsilon) > 0$ such that

$$\begin{aligned}\|T_1h - T_2h\| \le \|f(x_0+h)-f(x_0)-T_1h\| \\ +\|f(x_0+h)-f(x_0)-T_2h\| \le 2\epsilon\|h\|,\end{aligned}$$

provided that $\|h\| < \delta$. Divide both members of this inequality by $\|h\| \neq 0$, and then it follows from the linearity of T_1 and of T_2 that

$$\|T_1x - T_2x\| \le \epsilon, \ \forall x \in M, \ \|x\| = 1,$$

and thus $T_1 = T_2$.

In order to prove the continuity of f at a point x_0 where it happens to be differentiable, we use the triangle inequality:

$$\begin{aligned}\|f(x_0+h)-f(x_0)\| &\le \|f(x_0+h)-f(x_0')h\| \\ &+\|f'(x_0)h\| \le \epsilon\|h\|+\|f'(x_0)\|\cdot\|h\| \to 0.\end{aligned}$$

(*f*) It may be quickly verified that, if f and g are both differentiable at x_0, then $\alpha f+\beta g$ will also be so, for arbitrary real values α and β. Besides, the identity below holds

$$(\alpha f+\beta g)'(x_0)=\alpha f'(x_0)+\beta g'(x_0).$$

(*g*) When $f : M \to I\!R$ is differentiable at a point x_0 and $f'(x_0)=0$, then x_0 is said to be a stationary point, with the options of being a **local minimum**, a **local maximum**, or a **saddle point**. In optimization problems, the search for points of local minimum – or maximum – is thus strongly related to finding stationary points, as long as the model under study deals with differentiable functions. A remark worth to be made: it may happen that no internal point minimizes (or maximizes) globally the considered function, as such extreme values may be held at the domain boundary.

We state now the property known as the

Chain Rule *Take $M, N,$ and P as normed espaces where for the functions $f : M \to N, g : N \to P$ the existence of $f'(x_0)$ and $g'(f(x_0))$ is assured. Then the composite function $h := g \circ f$ is differentiable at x_0 and we have that*

$$h'(x_0)=g'(f(x_0))\circ f'(x_0).$$

Proof It suffices to estimate

$$\Delta := g(f(x_0+h))-g(f(x_0))-[g'(f(x_0))\circ f'(x_0)]h.$$

From the triangle inequality, besides adding and subtracting the same term, it follows that

$$\begin{aligned}\|\Delta\| \le \|&g(f(x_0+h))-g(f(x_0))-g'(f(x_0))[f(x_0+h)-f(x_0)]\| \\ &+\|g'(f(x_0))[f(x_0+h)-f(x_0)-f'(x_0)h]\|,\end{aligned}$$

and from it we are driven to $\|\Delta\|/\|h\| \to 0$, if $\|h\| \to 0$, as a consequence of the differentiability of g and of f, plus the continuity of f at x_0.

Example 7.1 For functions $f : D \subset I\!R^n \to I\!R^p$, as long as the operator $f'(x_0) : I\!R^n \to I\!R^p$ is linear, it is identified to a matrix. Such a matrix turns out to be exactly the Jacobian matrix

$$\left[\frac{\partial f_\imath}{\partial x_\jmath}(x_0)\right], 1 \leq \imath \leq p, 1 \leq \jmath \leq n.$$

Example 7.2 Take the nonlinear operator

$$\left.\begin{array}{l} F : C^0(0,1) \to C^1(0,1) \\ f \quad\; \to \int_0^t G(S, f(S))dS \end{array}\right],$$

for a given $G : I\!R^2 \to I\!R$. Suppose now that G is continuous and that $D_2 G$ is continuous as well.[2] It follows then, for each fixed $f_0 \in C^0(0,1)$ and for arbitrary $h \in C^0(0,1)$, that

$$\begin{aligned}[F(f_0 + h) - F(f_0)](t) &= \textstyle\int_0^t [G(S, f_0(S) + h(S)) - G(S, f_0(S))]dS \\ &= \textstyle\int_0^t [D_2 G(S, f_0(S) + \theta_S h(S))h(S)]dS.\end{aligned}$$

Therefore, since

$$\begin{aligned}&F(f_0 + h) - F(f_0) - \int_0^t D_2 G(S, f_0(S))h(S)dS \\ &= \int_0^t [D_2 G(S, f_0(S) + \theta_S h(S)) - D_2 G(S, f_0(S))]h(S)dS,\end{aligned}$$

being $D_2 G$ uniformly continuous for any compact, we deduce, for arbitrary values of $\epsilon > 0$, the existence of corresponding values of $\delta > 0$ for which

$$\begin{aligned}&\left|\int_0^t [D_2 G(S, f_0(S) + \theta_S h(S)) - D_2 G(S, f_0(s))]h(S)dS\right| \\ &\leq \left|\int_0^t \epsilon h(S)dS\right| \leq \|h\|_\infty \epsilon \int_0^t dS \leq \epsilon \|h\|_\infty,\end{aligned}$$

provided $\|h\|_\infty < \delta$.

As a conclusion,

$$\left[F'(f_0)h\right](t) = \int_0^t D_2 G(S, f_0(S))h(S)dS.$$

[2] $D_2 G$: notation for the partial derivative $\partial G/\partial x_2$, following Example 7.1.

7.2 Mean Value Theorems

This section main task is to present an important tool for deducing inequalities, particularly the so-called **a priori** estimates , that hold a fundamental status within the search for the convergence of numerical schemes.

Consider a function $f : M \to I\!R$ assumed to be differentiable at every point of the segment from x_1 to x_0, contained in its domain. In other words, suppose to be defined the mapping

$$\begin{array}{ll} f' : [x_0, x_1] \to \mathcal{L}(M, I\!R) \\ \quad x \quad \to f'(x) \end{array}.$$

(With $\mathcal{L}(M, N)$ it is denoted the set of all linear continuous transformations from M to N, here chosen $N = I\!R$.)

Under these conditions,

$$\begin{array}{ll} \psi : [0, 1] \to I\!R \\ \quad \lambda \quad \to \psi(\lambda) := f(x_0 + \lambda(x_1 - x_0)) \end{array}$$

is differentiable on $[0, 1]$, since $\psi = f \circ g$, being

$$\begin{array}{ll} g : [0, 1] \to M \\ \quad \lambda \quad \to g(\lambda) := x_0 + \lambda(x_1 - x_0) \end{array}$$

differentiable, due to (b), (c) and (f). Applying the Mean Value Theorem (from the Differential Calculus on the line) to ψ, one concludes the existence of a point $\overline{x}$ on the segment (x_0, x_1), for which

$$f'(\overline{x})(x_1 - x_0) = f(x_1) - f(x_0). \tag{7.3}$$

To reach (7.3) we have made use of the chain rule for ψ while, for g, the identification of operators from $\mathcal{L}(I\!R, M)$ to vectors from M.

We have thus concluded the proof of the Mean Value Theorem for real functions with domain in an arbitrary normed space. It is known that this result cannot be extended to vector-valued functions, even to $N = I\!R^2$. Through an intuitive reasoning, we cannot have such an extension because, as any f is "split in its components," it is natural to expect generating different domain points $\overline{x}$ for each component. Nevertheless, it is held as valid the following:

Theorem 7.1 (Mean Value Inequality) *Let $f : M \to N$ be differentiable on the segment $[x_0, x_1] \subset M$. It is then guaranteed to exist $\theta \in (0, 1)$ and $\bar{x} \in M$, being $\bar{x} := x_0 + \theta(x_1 - x_0)$, such that*

$$\|f(x_1) - f(x_0)\| \leq \|x_1 - x_0\| \cdot \|f'(\bar{x})\|,$$

or, otherwise said:

$$\|f(x_1) - f(x_0)\| \leq \|x_1 - x_0\| \sup_{\theta \in (0,1)} \|f'(x_0) + \theta(x_1 - x_0))\|. \tag{7.4}$$

Proof Let $\tilde{\ell} \in N^*$ be an arbitrary functional. Then $g := \tilde{\ell} f : M \to I\!R$ is differentiable in $[x_0, x_1]$ due to the chain rule, and it is verified that

$$g'(x) = \tilde{\ell} f'(x).$$

The just proved version for the real line of the Mean Value Theorem implies that

$$\|f(x_1) - f(x_0)\| = (\tilde{\ell} \circ f'(\bar{x}))(x_1 - x_0) \tag{7.5}$$

for some $\bar{x} \in [x_0, x_1]$, and such $\bar{x}$ depends on the functional $\tilde{\ell}$. (The identity (7.5) may be seen as a weak formulation of the Mean Value Theorem.)

Hahn-Banach Theorem implies the existence of a functional $\ell \in N^*$, with $\|\ell\| = 1$ and for which

$$\|f(x_1) - f(x_0)\| = \ell(f(x_1) - f(x_0)).$$

By making use of (7.5), it follows that

$$\|f(x_1) - f(x_0)\| \leq \|\ell\| \|f'(\bar{x})(x_1 - x_0)\| = \|f'(\bar{x})(x_1 - x_0)\| \leq$$
$$\leq \|f'(\bar{x})\| \cdot \|x_1 - x_0\|,$$

since $\|\ell\| = 1$, which closes the proof.

Hereby a simple consequence of the mean value inequality:

Whenever $f'(x) = 0$, for every x in a connected open subset A of M, then f is constant all over A.

7.3 Higher-Order Derivatives

When a function $f : M \to N$ turns out to be differentiable all over its domain, a function g becomes naturally defined:

$$\left.\begin{array}{rl} g := f' : M & \to \mathcal{L}(M, N) \\ x & \to f'(x) \end{array}\right].$$

One may wonder if such g is differentiable at $x_0 \in M$. If so, it is said that f is twice differentiable at x_0, denoting then this derivative by $f''(x_0) \in \mathcal{L}(M, \mathcal{L}(M, N))$.

Let us have a look at the elements $\tilde{B}$ of $\mathcal{L}(M, \mathcal{L}(M, N))$. For a given $x \in M$, we have that $\tilde{B}(x) \in \mathcal{L}(M, N)$; this way, given any $y \in M$, it follows that $\tilde{B}(x)y \in N$. Observe that $\tilde{B}(x)y$ is linear at x, for fixed y, as well as on y, for x fixed; in other words, the function $(x, y) \mapsto \tilde{B}(x)y$ is **bilinear**.

This fact allows to identify in a natural way the space $\mathcal{L}(M, \mathcal{L}(M, N))$ to the space $\beta(M, N)$ of the bounded bilinear mappings from M to N, more properly, from $M \times M$ to N. (Let us recall, cf. Sect. 3.3.1, that a bilinear transformation $B \in \beta(M, N)$ is said to be bounded when

$$\sup_{\|x\|_M = \|y\|_M = 1} \|B(x, y)\|_N < \infty.$$

The value of this *supremum* defines the norm of B.)

The identification

$$\left.\begin{array}{lllll} \mathcal{I} : \mathcal{L}(M, \mathcal{L}(M, N)) \to \beta(M, N) & & & & \\ \qquad\quad \tilde{B} & \to & B: & M \times M \to & N \\ & & & (x, y) & \to B(x, y) := \tilde{B}(x)y \end{array}\right]$$

is linear, 1-1 and onto. Since

$$\|B(x, y)\| = \|\tilde{B}(x)y\| \leq \|\tilde{B}(x)\|\|y\| \leq \|\tilde{B}\|\|x\|\|y\|, \tag{7.6}$$

this is a continuous transformation. When N is complete, then $\beta(M, N)$ and $\mathcal{L}(M, \mathcal{L}(M, N))$ are both Banach spaces, then it may be claimed that this mapping has a continuous inverse, as a consequence of Banach Isomorphisms theorem.

Indeed, the continuity of the inverse $\mathcal{I}^{-1}$ may be concluded directly, and, moreover, we may deduce the equality $\|\tilde{B}\| = \|B\|$ even when M and N lack completeness. Given $B \in \beta(M, N)$, being $\tilde{B}(x)y = B(x, y)$, *i.e.*, $\tilde{B} = \mathcal{I}^{-1}(B)$, it holds, for fixed x:

$$\|\tilde{B}(x)y\| \leq \|B\|\|x\|\|y\|,$$

or

$$\|\tilde{B}(x)\| \leq \|B\|\|y\|,$$

and from this it follows

$$\|\tilde{B}\| \leq \|B\|. \tag{7.6'}$$

The inequalities (7.6) and (7.6′) imply the equality

$$\|\tilde{B}\| = \|B\|.$$

It becomes more comfortable and natural to think on the second derivative of f as an element of $\beta(M, N)$.

Now let $f : M \to \mathbb{R}$ be a real valued function and x_0 be a point of minimum for f. It is known that $f'(x_0) = 0$ and that the bilinear form $f''(x_0)$ is positive, which means

$$f''(x_0)(x, x) > 0, \forall x \in M, x \neq 0.$$

When $M := \mathbb{R}^p$ and $N := \mathbb{R}$, the bilinear transformation $f''(x_0)$ is identified to the so-called Hessian matrix

$$\left[\frac{\partial^2 f}{\partial x_i \partial x_j}(x_0)\right]_{1 \leq i, j \leq p}.$$

For the same token, if $f''(x)$ exists for every x in a particular open set A in M, we have a mapping

$$f'' : A \subset M \to \beta(M, N).$$

As long as the derivative of that mapping at the point $x_0 \in A$ exists, it is named the derivative of order 3 for f at x_0, denoted by $f'''(x_0)$. It is then an element of $\mathcal{L}(M, \beta(M, N))$ – or of $\mathcal{L}(M, \mathcal{L}(M, N))$. By the same track taken above, we shall identify

$$\mathcal{L}(M, \beta(M, N)) \Longleftrightarrow \mathcal{T}(M, N),$$

where $\mathcal{T}(M, N)$ denotes the set of the mappings $T : M \times M \times M \to N$, that hold the property of being linear as regards to each coordinate **separately**, being called then trilinear.

We follow thus the same pattern, so that the derivative of order k for f at x_0 is denoted by $f^{(k)}(x_0)$. This happens to be then a k-linear mapping, that is, **multilinear** of order k from M to N, or once more, from the product of k copies of M to N.

It suffices to follow the same order of ideas presented for the Mean Value Theorem in order to prove, for $f : M \to \mathbb{R}$, that

$$f(x_0 + h) = f(x_0) + f'(x_0)h + f''(x_0 + \theta h)(h, h)/2! \tag{7.7}$$

for some θ, with $0 < \theta < 1$, provided there exist $f'(x)$ and $f''(x)$ for $x \in \{x_0 + \lambda h, 0 \leq \lambda \leq 1\}$.

More generally, under the corresponding hypotheses, by denoting

$$h^{(k)} := (h, h, \ldots, h) \in M^k,$$

we shall have

$$f(x_0+h) = f(x_0) + f'(x_0)h + f''(x_0)h^{(2)}/2! \\ +f'''(x_0)h^{(3)}/3! + \ldots + f^{(k)}(x_0+\theta h)h^{(k)}/k!, \tag{7.7'}$$

which is the k-th order Taylor formula.

For $f : M \to N$, the mean value inequality may be extended under the expression

$$\|f(x_0+h) - f(x_0) - f'(x_0)h\| \leq \\ \|f''(x_0+\theta h)(h,h)\|/2 \leq \|f''(x_0+\theta h)\| \|h\|^2/2 \tag{7.8}$$

or, in a more general estimate,

$$\|f(x_0+h) - f(x_0) - f'(x_0)h - \ldots - f^{(k-1)}(x_0)h^{(k-1)}/(k-1)!\| \\ = \|f^{(k)}(x_0+\theta h)h^{(k)}/k!\| \leq \|f^{(k)}(x_0+\theta h)\| \|h\|^k/k!, \tag{7.8'}$$

based on Taylor expansion of order k.

Exercise 7.1 Let $f \in C_0^k(\Omega)$ be a real function, where $\Omega \subset I\!R^n$ is a bounded open set. Equivalence holds for the norms

$$\|f\|_{k,2} := \left[\sum_{|\ell|=0}^{k} \|D^\ell f\|_0^2\right]^{1/2}, \; |||f|||_k := \left[\sum_{J=0}^{k} |||f^{(J)}|||_0^2\right]^{1/2}.$$

Here, $|||T|||$ denotes the norm of the multilinear mapping T.

Hint: it is enough to prove that $\sum_{|\ell|=J} \|D^\ell f\|_0^2$ and $|||f^{(J)}|||_0^2$ are equivalent semi-norms. It occurs that, being $\{e_m\}_{1\leq m\leq n}$ the vectors from the canonical basis of $I\!R^n$, namely, $e_m := (\delta_m^p)_{1\leq p\leq n}$ and $\ell = (\ell_1, \ell_2, \ldots, \ell_n)$ a multi-index, with $|\ell| = J$, we have

$$D^\ell f(x) = f^{(J)}(x) \cdot \xi,$$

where

$$\xi := (\underbrace{e_1, \ldots, e_1}_{\ell_1}, \underbrace{e_2, \ldots, e_2}_{\ell_2}, \ldots, \underbrace{e_n, \ldots, e_n}_{\ell_n}). \tag{7.9}$$

From this remark, from the equivalence of the norms in $I\!R^S$ and from being $[I\!R^n]^J$ generated by the vectors ξ in the form (7.9), we reach the conclusion for the proof of the announced result.

7.4 Iterative Methods

Almost surely the most familiar iterative method in Analysis is the one described by the

Fixed Point Theorem (Banach) *Suppose to be $f : M \to M$ a* ***contraction*** *on the complete metric space M, which means: being d the distance in M, there exists a constant $\rho \in]0, 1[$ for which*

$$d(f(x_1), f(x_2)) \leq \rho d(x_1, x_2), \forall x_1, x_2 \in M. \tag{7.10}$$

Then, under such hypotheses, there exists a unique solution for the equation

$$f(x) = x. \tag{7.11}$$

In other words, the function f has one and only one ***fixed point****.*

This fixed point $\bar{x}$ may be reached throughout the following ***iteration scheme****:*

$$x_{j+1} := f(x_j) \text{ for } j \geq 0; x_0 \text{ arbitrary }, \tag{7.12}$$

since the sequence (x_j) in (7.12) converges to $\bar{x}$ and, moreover

$$d(x_j, \bar{x}) \leq \frac{\rho^{j+1}}{1-\rho} d(x_1, x_0). \tag{7.13}$$

Therefore,

$$d(x_{j+1}, \bar{x}) \leq C d(x_j, \bar{x}), C := d(x_0, x_1), \tag{7.14}$$

that is, the convergence of scheme (7.12) is ***linear****.*

The proof for this important result is simple indeed and shorter than its whole statement: enough to verify, cf. [16], that (7.12) defines a Cauchy sequence, which may be concluded by looking at the sum

$$d(x_{N+p}, x_N) \leq \sum_{i=N}^{N+p-1} d(x_{i+1}, x_i).$$

Example 7.3 Consider $M := [1, \infty)$ and

$$\left.\begin{array}{l} f : M \to M \\ \quad x \to x + 1/x \end{array}\right] .$$

For this function f, (7.11) presents no solution. Nevertheless, it occurs that, for some $\xi := x + \theta(y - x), 0 < \theta < 1$, the relation

$$f(x) - f(y) = f'(\xi)(x - y) = (1 - 1/\xi^2)(x - y)$$

is verified, and therefore

$$|f(x) - f(y)| < |x - y|,$$

which means that (7.10) holds for $p = 1$.

Example 7.3 illustrates the need of being the constant ρ in (7.10) **strictly** smaller than 1.

Exercise 7.2 Verify that the same conclusion of the above theorem holds, as long as f fulfills its hypotheses with the exception of (7.10), which is then replaced by

$$f^n := f \circ f^{n-1}, \text{ for some } n > 1.$$

As long as one makes use of the Mean Value Theorem – it should be said, the mean value inequality – it can be stated the

Theorem 7.2 *Take B as a Banach space and let $f : B \to B$ be a differentiable function in B, such that*

$$\sup_{x \in B} \|f'(x)\| < 1. \tag{7.15}$$

Then f is a contraction and thus for it holds the Fixed Point Theorem.

Example 7.4 Let us turn back to Example 7.2. Being the derivative of

$$F : f \to \int_0^t G(S, f(S))dS$$

given by

$$[F'(f_0)h](t) := \int_0^t D_2G(S, f_0(S)) \cdot h(S)dS,$$

for every f_0 which fulfills

$$\sup_{0 \leq S \leq t} |f_0(S)| \leq \alpha,$$

it follows that

$$|[F'(f_0)h](t)| \leq t \sup_{0\leq S\leq t} |h(S)| \sup_{\substack{0\leq S\leq t\\ -\alpha\leq\sigma\leq\alpha}} |D_2G(S,\sigma)|.$$

Thus, for t small enough,

$$\|F'(f)\| \leq \rho < 1$$

holds, and this assures the existence proof for the solution of a first-order ordinary differential equation.

To study solutions to nonlinear algebraic equations for real functions of real variables,

$$F(\bar{x}) = 0, \tag{7.16}$$

one ought[3] to appeal to iterative methods. Among these, one which beats a lot of its counterparts is the **Newton-Raphson** method:

$$x_{j+1} := x_j - \frac{F(x_j)}{F'(x_j)},\ j \geq 0;\ x_0 \text{ given.} \tag{7.17}$$

When it converges, this method exhibits the property

$$|x_{j+1} - \bar{x}| \leq C|x_j - \bar{x}|^2, \tag{7.18}$$

for some constant C, which means that its **convergence** order is **quadratic** – compare (7.18) with (7.14).

To make use of (7.17), it becomes a real need that F to be a differentiable function and that $F'(x_j) \neq 0$. Observe that, being $F'(x) \neq 0$, (7.17) is precisely (7.12) applied to

$$f(x) := x - F(x)/F'(x),$$

because, for f thus defined, (7.16) is equivalent to (7.11).

Observe further that this approximation algorithm has already been employed. Its job has been done while we were dealing with the construction of the example quoted as Encounter 5, Sect. 1.1.

In order to apply Theorem 7.2 to f, suppose that $F \in C^2(I\!R)$, besides being $F'(\bar{x}) \neq 0$. It is thus deduced:

[3] Observe that the polynomial equations live among those problems in the solving queue. A classical result from Algebra would not issue a permit to discover a finite number of steps algorithm fit to solve all elements in this above problem collection. That is why we are lead to the iteration ring. Remember that the eigenvalue search also enjoys this environment.

$$f'(x) = F(x)F''(x)/F'(x)^2,$$

and from that one concludes, for $x \to \bar{x}$, $f'(x) \to 0$. Therefore, for points that lie "near enough" the root $\bar{x}$, the auxiliar function f turns out to be a contraction, and thus (7.17) defines a convergent sequence to $\bar{x}$.

Our goal now is to generalize (7.17) for vector functions, duly told as

Newton-Raphson Vector Method Let be $\mathcal{F} : B \to B$ a function on a Banach space B. Suppose $\mathcal{F} \in C^2(B)$, while looking for approximated solutions to

$$\mathcal{F}(\bar{x}) = 0.$$

Make then the hypothesis that $\mathcal{F}'(\bar{x})$ is invertible, with $[\mathcal{F}'(\bar{x})]^{-1}$ bounded. In this setting, the iteration scheme

$$x_{j+1} := x_j - [\mathcal{F}'(x_j)]^{-1}\mathcal{F}(x_j),\ j \geq 0;\ x_0 \text{ given}, \tag{7.17'}$$

converges to $\bar{x}$, with a quadratic convergence order.

For example, suppose that $\tilde{F} : I\!R^p \to I\!R^p$ satisfies the hypotheses above, being $\tilde{F} := (F_1, \ldots, F_p)$ and $\tilde{x} := (x^1, \ldots, x^p)^t \in I\!R^p$, where the transpose of $(x^1, \ldots, x^p)$ is denoted by $(x^1, \ldots, x^p)^t$.

Newton scheme takes then the form

$$\tilde{x}_{j+1} := \tilde{x}_j - (D_k F_\ell(\tilde{x}_j))^{-1}\tilde{F}(\tilde{x}_j),$$

where

$$(D_k F_\ell) = \left(\partial F_\ell / \partial x^k\right)_{1 \leq \ell, k \leq p}$$

is the Jacobian matrix for $\tilde{F}$.

Quite often, the modified scheme

$$x_{j+1} := x_j - [\mathcal{F}'(x_0)]^{-1}\mathcal{F}(x_j),\ j \geq 0;\ x_0 \text{ given}, \tag{7.17''}$$

replaces (7.17′). This strategy bears the property of liberating the user from the inversion of $\mathcal{F}'(x_j)$ on every iteration.

Other options are to get hold of the so-called quasi-Newton schemes, for which the inverse of the derivative, $[\mathcal{F}'(x_j)]^{-1}$, is **simulated** at each iteration by a conveniently defined operator H_k. One is then lead to

$$x_{j+1} := x_j - H_k\mathcal{F}(x_j),\ j \geq 0;\ x_0 \text{ given}, \tag{7.17'''}$$

being the curious reader invited to browse, for example, on [28].

Chapter 8
Hilbert Bases and Approximations

8.1 Orthogonalization

This chapter works with a real Hilbert space H whose definition is based on an inner product $(\cdot|\cdot)$.

A subset $S \subset H$ is said to be **orthogonal** if its elements are pairwise orthogonal, which means

$$s_1, s_2 \in S, s_1 \neq s_2 \Longrightarrow (s_1|s_2) = 0.$$

An orthogonal set is said to be **orthonormal** if all its elements have norm equal to 1. This way, a set $S := \{s_\iota\}$ is orthonormal if and only if

$$(s_\iota|s_J) = \delta_{\iota J} (\delta_{\iota J} := \text{ Kronecker delta}).$$

An orthonormal set is shown to be always linearly independent. Indeed, suppose that

$$\sum_{J=1}^{N} \alpha^J s_{\iota_J} = 0, \tag{8.1}$$

for some **finite** collection of real numbers $\{\alpha^J\}, J = 1, \ldots, N$ and vectors $\{s_{\iota_J}\}_{J=1,\ldots,N} \subset S$. It follows then, from multiplication of (8.1) by each one of the vectors s_{ι_K}, that

$$\alpha^K (s_{\iota_K}|s_{\iota_K}) = \alpha^K \|s_{\iota_K}\|^2 = \alpha^K = 0.$$

C. A. de Moura, *Functional Analysis Tools for Practical Use in Sciences and Engineering*, https://doi.org/10.1007/978-3-031-10598-2_8

Conversely, given a sequence $\{s_\jmath\}_{\jmath\in I\!N}$ of linearly independent vectors, it is possible to reach, throughout this very sequence, another one $\{\sigma_\jmath\}_{\jmath\in I\!N}$, which shows up to be orthogonal and which **generates** the same subespace.[1]

The just stated converse property may be verified by the following process, known as the **Gram-Schmidt orthonormalization**.

Define

$$\sigma_1 := s_1/\|s_1\|, \tag{8.2}$$

which can be carried over, since $s_1 \neq 0$, as $\{s_\jmath\}$ are all linearly independent. Let now be

$$\tilde{s}_2 := s_2 - (s_2|\sigma_1)\sigma_1. \tag{8.3}$$

It is seen that $(\tilde{s}_2|\sigma_1) = 0$ and further $\tilde{s}_2 \neq 0$; we then assign

$$\sigma_2 := \tilde{s}_2/\|\tilde{s}_2\|. \tag{8.2/}$$

(Observe that in (8.3) it has been defined $\tilde{s}_2$ in such a way as to eliminate from s_2 its projection on the direction of σ_1.) Recursively define, for $n \geq 2$,

$$\tilde{s}_n := s_n - \sum_{\jmath=1}^{n-1}(s_n|\sigma_\jmath)\sigma_\jmath, \tag{8.3'}$$

$$\sigma_n := s_n/\|s_n\|. \tag{8.2''}$$

The set $\{\sigma_\jmath\}$ exhibits thus the announced properties.

Exercise 8.1 (Modified Gram-Schmidt process) The procedure contained in (8.2)–(8.3) may be replaced, with computational advantages,[2] reordering the calculations according to the algorithm described in the sequel.

Verify that, when **exactly** performed these calculations, (*i.e.*, with whole precision, no rounding errors admitted), we are lead to the same orthonormal set, in both processes:

---

[1] A set $A \subset H$ is said to **generate** the subspace $[A]$, or that $[A]$ is the subspace generated by A, if any vector $v \in [A]$ may be described as

$$v = \sum_{\imath\in I}\alpha^\imath a_\imath,$$

for some finite set of vectors $a_\imath \in A$ and of real numbers $\alpha^\imath$.

[2] we mean, in such a way as to assure what is called *computational stability*.

```
For k = 1,...,n, perform
      p(k) := k
end
For k = 1,...,n, perform
    Atribute to the variable J the smallest value of
    J = k,...,n  for which
          ||s_p(J)|| ≥ ||s_p(ι)|| ∀ι = k,...,n
    σ_k := s_p(J)/||s_p(J)||
    For ι = k,..., J − 1 perform
          p(ι + 1) := p(ι)
    end
    For ι = k + 1,...,n, perform
          s_p(ι) := s_p(J) − (s_p(J)|σ_k)σ_k
    end
end
```

It is worth remarking that, in H, given $S := \{s_\iota\}_{\iota\in\mathbb{N}}$ – explicitly, a countable non-empty set – it is possible to extract from itself a linearly independent subset $\mathcal{S}$, such that $[\mathcal{S}] = [S]$.

Indeed, suppose $S \neq \{0\}$ and define recursively the subsequence s_{ι_k}:

(*a*) $\iota_1 := \min\{\jmath \in \mathbb{N};\, s_\jmath \neq 0\}$;
(*b*) defined $\iota_1, \iota_2, \ldots, \iota_n$, let $V_n := [\{s_{\iota_\jmath};\, 1 \leq \jmath \leq n\}]$, then

$$\iota_{n+1} := \min\{\jmath \in \mathbb{N};\, s_{\iota_\jmath} \notin V_n\}.$$

(Whenever $\{\jmath \in \mathbb{N};\, s_{\iota_\jmath} \notin V_n\} = \emptyset$, it is enough to define

$$\mathcal{S} := \{s_{\iota_1}, s_{\iota_2}, \ldots, s_{\iota_n}\}.)$$

8.2 Fourier Series

An orthonormal set $\{e_\jmath\}$ of vectors from H is called a **complete system**, or a **Hilbert basis**, if $\{e_\jmath\}^\perp = \{0\}$; otherwise said, the only vector from H to be orthogonal to all vectors[3] $\{e_\jmath\}$ is the null vector.

Our scope now is to prove that, given a complete system $\{e_\jmath\}_{\jmath\in\mathbb{N}}$, any vector $x \in H$ may be described in the format

[3] We will herewith always suppose $\{e_\jmath\}$ countable!

$$x = \sum_{j=1}^{\infty} x^j e_j \tag{8.4}$$

for a convenient choice of the real numbers x^j. But, beforehand, there is the need to explain what is meant with a series like (8.4).

Given the vectors $\{v_j\}$ from H, the **series** $\sum_{j=1}^{\infty} v_j$ is said to be **convergent** with $v_0 \in H$ as its limit, so the writing

$$v_0 = \sum_{j=1}^{\infty} v_j$$

if

$$v_0 = \lim_{N\to\infty} \sum_{j=1}^{N} v_j.$$

Suppose then that such an expansion in the form (8.4) will be possible for the vector x, and let $x_N := \sum_{j=1}^{N} x^j e_j$. Since $x_N \to x$, due to (iii), Sect. 2.4, $(x_N|e_J) \to (x|e_j)$, for every fixed $j \in \mathbb{N}$.

Now, since $j < N$,

$$(x_N|e_j) = \sum_{j=1}^{N} x^j \delta_{ij} = x^j.$$

As a consequence,

$$x^j = \lim_{N\to\infty} (x_N|e_j) \to (x|e_j),$$

which means that, if an expansion on the form (8.4) exists, the real numbers x^j are determined uniquely:

$$x^j = (x|e_j),\ j = 1, 2, \ldots. \tag{8.5}$$

It is then said that the series in the right-hand side of (8.4), with the scalar coefficients x^j defined in (8.5), is the **Fourier expansion** of the vector x with respect to the basis $\{e_j\}$. The scalar numbers x^j are called **Fourier coefficients** of x for the basis $\{e_j\}$: they are, except for the sign, the absolute value of the projections of x through the directions of e_j.

The next step is to prove that the expansion (8.4) effectively holds, as long as x^j be defined by (8.5).

Let $S_N := [\{e_j;\ j \leq N\}]$. It is already known that S_N is a closed subspace (Exercise 3.4a). Thus, the orthogonal projection on S_N exists and will be denoted by $P_N : H \to S_N$. It is at once deduced that

$$P_N x = \sum_{j=1}^{N} (x^j | e_j). \tag{8.6}$$

(In fact, as observed in part (b) of Exercise 3.4, Sect. 3.4, for a finite dimension subspace, it is possible to prove directly the Projection Theorem, that is, to show that (8.6) defines the orthogonal projection S_N.)

As observed on Sect. 3.4, $\|P_N\| = 1$, thus $\|P_N x\| \leq \|x\|$, and from (8.6) it follows

$$\|P_N x\|^2 = \sum_{j=1}^{N} |x^j|^2 \leq \|x\|^2,$$

which implies the so-called Bessel inequality:

$$\sum_{j=1}^{\infty} |x^j|^2 \leq \|x\|^2, \forall x \in H. \tag{8.7}$$

Observe that (8.7) holds for any orthonormal system $\{e_j\}$, independently of being it complete.

From Bessel inequality it follows that $\{x^j\} \in \ell^2$ and, moreover, that the series (8.4) converges, since $\{x_N\}$ is a Cauchy sequence:

$$\|x_{N+P} - x_N\|^2 = \left\| \sum_{j=N+1}^{N+P} x^j e_j \right\|^2 = \sum_{j=N+1}^{N+P} |x|^2 \to 0, \text{ if } N \to \infty, \forall P.$$

Therefore, there exists $x_0 \in H$ for which

$$x_N \to x_0.$$

It remains to be shown that $x_0 = x$. Now, we may use the same track that has led to (8.5), so as to get

$$(x_0 | e_j) = x_j, \tag{8.5'}$$

from which it follows

$$(x - x_0 | e_j) = 0, \forall j \in I\!N.$$

We use then the fact of being $\{e_j\}$ a complete system in order to conclude that $x_0 = x$ and, thus, (8.4) holds.

Given two vectors $x, y \in H$, again thanks to the continuity of the scalar product – cf. (2.13.c) – it follows that

$$\begin{aligned}(x|y) &= \left(\sum_{j=1}^{\infty}(x|e_j)e_j \,|\, \sum_{k=1}^{\infty}(y|e_k)e_k\right)\\ &= \lim_{N\to\infty}\left(\sum_{j=1}^{N}(x|e_j)e_j \,|\, \sum_{k=1}^{N}(y|e_k)e_k\right)\\ &= \lim_{N\to\infty}\sum_{j=1}^{N}\sum_{k=1}^{N}(x|e_j)(y|e_k)(e_j|e_k)\\ &= \lim_{N\to\infty}\sum_{j=1}^{N}(x|e_j)(y|e_j) = \sum_{j=1}^{\infty}(x|e_j)(y|e_j).\end{aligned}$$

In plain words, we have

$$(x|y) = \sum_{j=1}^{\infty}(x|e_j)(y|e_j). \tag{8.8}$$

From this equality, we deduce in particular the **Parseval identity**: by taking $y := x$, it follows

$$\|x\|^2 = \sum_{j=1}^{\infty}|(x|e_j)|^2. \tag{8.9}$$

We can think that such identity generalizes Pythagoras Theorem.

Observe the fact that (8.4) to hold for every $x \in H$ implies $[\{e_j\}]$ to be dense in H. Conversely, if $[\{e_j\}]$ is dense in H, the orthonormal system $\{e_j\}$ is necessarily complete.

Indeed, being P_N the orthogonal projection on S_N as defined in (8.6), it is true, for any collection of real values $\alpha^1, \ldots, \alpha^N$,

$$\|x - P_N x\|^2 \leq \|x - \sum_{j=1}^{N}\alpha^j e_j\|^2.$$

Through another saying, the Fourier coefficients provide the **best approximation** which is available for x in S_N, with respect to, of course, the norm from H. This implies that, as long as we could approximate the vectors from H by finite linear combinations of the elements e_j's, necessarily we will have $x_n \to 0$, which means that (8.4) will hold.

An orthonormal system $\{e_j\}$ is said to be **maximal** if another set which properly contains $\{e_j\}$ fails to be orthonormal.

We can state the following

Theorem Let H be a Hilbert space and $\{e_j\}$ be an orthonormal system. All the following conditions about $\{e_j\}_{j\in\mathbb{N}}$ turn out to be equivalent:

(i) $\{e_\jmath\}$ is maximal.
(ii) $\{e_\jmath\}$ is complete (*i.e.*, $(x|e_\jmath) = 0, \forall \jmath \in I\!N \Longrightarrow x = 0$).
(iii) $x = \sum_{\jmath=1}^{\infty} (x|e_\jmath)e_\jmath, \forall x \in H$.
(iv) $[\{e_\jmath\}]$ is dense in H.
(v) $(x|y) = \sum_{\jmath=1}^{\infty} (x|e_\jmath)(y|e_\jmath), \forall x, y \in H$.
(vi) $\|x\|^2 = \sum_{\jmath=1}^{\infty} (x|e_\jmath)^2, \forall x \in H$.

The proof for $(ii) \Longrightarrow (iii) \Longrightarrow (iv) \Longrightarrow (v) \Longrightarrow (vi)$ has already been presented.

Suppose now that $x_0 \neq 0$, but $(x_0|e_\jmath) = 0, \forall \jmath \in I\!N$. Then $\{e_\jmath\}$ must be a proper subset of $\{x_0/\|x_0\|\} \cup \{e_\jmath\}$, which is orthonomal. This shows that $(i) \Longrightarrow (ii)$ holds.

In an analogous fashion, if $\{e_\jmath\}$ is not maximal, there must exist $\{x_0\} \cup \{e_\jmath\}$ which is orthonormal, and this implies

$$0 = \sum_{\jmath=1}^{\infty} |(x_0|e_\jmath)|^2 = \|x_0\|^2 \neq 0.$$

That shows thus $(vi) \Longrightarrow (i)$.

It may be recalled at this point that there exists another concept of basis for infinite dimensional vector spaces: a subset β of a vector space V is said to be a **Hamel basis** for V if, for arbitrary $v \in V$, there exist finite subsets $\{b_\iota\}_{\iota \in I} \subset \beta$ and $\{\alpha^\iota\}_{\iota \in I} \subset I\!R$ such that

$$v = \sum_{\iota \in I} \alpha^\iota b_\iota .$$

Besides, such a representation ought to be unique.[4]

8.3 Separable Spaces: Approximation

It is quite natural to pose the question: which are the conditions to assure existence in H of a Hilbert basis? Observe that, whenever it contains such a basis, the considered space is necessarily separable. As a matter of fact, the set of all finite linear combinations of elements from $\{e_j\}$, with rational coefficients, is countable. The converse is also true, and thus we have the

Theorem 8.1 *A Hilbert space is separable if and only if it contains an at most countable Hilbert basis.*

[4] The existence proof of such a basis, for any vector space, asks for help from Zorn Lemma, or, which is equivalent, of the Axiom of Choice, see [55].

Proof Let S be a countable dense set in H. It allows, then, to obtain from its elements a (necessarily) countable and linearly independent subset S_1, with the property $[S] = [S_1]$.

Gram-Schmidt orthonormalization process would then lead us to an orthonormal set S_2, always countable (as well dense in $\{x \in H;\ \|x\| = 1\}$), such that $[S_2] = [S]$. We claim that completeness for S_2 holds.

In fact, suppose that there exists $x_0 \in H$ such that $(x_0|s_k) = 0 \forall s_k \in S_2$. Being $[S] = [S_2]$ dense in H, there exists a sequence $\{x_N\}$ for which $x_N \to x_0$, where each x_N is a linear combination of elements $s \in S_2$

$$x_N := \sum_{J=1}^{M} \alpha_N^J s_N^J, \text{ with } M = M(N).$$

Once known that $(x_N|x_0) = 0, \forall N \in \mathbb{N}$ and $(x_N|x_0) \to \|x_0\|^2$, we may conclude that $x_0 = 0$, and thus S_2 is complete.

Exercise 8.2 Every separable Hilbert space is **isometrically isomorphic** to the space of sequences ℓ_2. In other words, there exists

$$\mathcal{J} : \ell_2 \to H,$$

linear, 1-1 and onto, for which

$$(\mathcal{J}x|\mathcal{J}y)_H = (x|y)_{\ell_2}, \forall x, y \in \ell_2.$$

Example 8.1 The space $L^2(0, 2\pi)$, being the completion of $C^0(0, 2\pi)$, is separable, since by Exercise 2.15, the latter is a separable space. The set

$$\beta := \{f \equiv 1\} \cup \{\cos nx,\ \sin nx\}_{n \in \mathbb{N}}$$

is orthogonal in $L^2(0, 2\pi)$. It may be verified that it is complete, as a consequence of the following version of

Weierstrass Approximation Theorem *Given $f \in C^0(0, 2\pi)$ and being $\epsilon > 0$, there exists a trigonometric polynomial $p = p_{\epsilon,f}$,*

$$p(x) := \sum_{k=1}^{N(\epsilon,f)} \{a_k \cos kx + b_k \sin kx\},$$

for which it holds

$$\|p - f\|_\infty < \epsilon.$$

According to this theorem, the closure of $[\beta]$ in $C^0(0, 2\pi)$, with the norm $\|\bullet\|_\infty$, contains $C_0^0(0, 2\pi)$. Therefore, the closure of $[\beta]$ in $L^2(0, 2\pi)$ contains $C_0^0(0, 2\pi)$, being, consequently, the same space $L^2(0, 2\pi)$.

Example 8.2 One may find some frameworks that would call for a special type of **approximation spaces** in H: these are finite dimensional subspaces $V_n \subset H$, such that $\cup_{n=1}^{\infty} V_n$ is dense in H. A possible choice for those spaces is $V_N := [\{e_j, 1 \leq j \leq N\}]$.

Example 8.3 The spaces $H^k(\Omega)$ are also separable, provided Ω fulfills some particular regularity conditions. A way to prove this fact is by building the approximation spaces of **finite elements**, like in [5], the particular example of $H^1(0, 1)$ being described in the sequel.

8.3.1 An Example: The Finite Elements

On Sect. 5.5.1 we have mentioned approximations numerically developed for the solution of a given differential equation. The quoted result, when deduced, looked for finite difference algorithms – which were those available to the authors [20] at that time. This solution method sets itself on the **pointwise** (or *strong*) **formulation** of the associated problem – initial value, boundary value, or mixed. Alternatively, the finite element method, which will be quite shortly summarized in the sequel, is **global**, as it makes use of the **variational formulation** (called *weak* as well). It is quite important to emphasize that, despite being often this last formulation considered as a mathematical gimmick, it turns out to be the most natural approach to several problems. This can be stated because it reflects the global modelling of particular behavior laws. The pointwise formulation – whose deduction, by passing from space regions with defined volumes to other regions with "arbitrarily small" volumes – incorporates an asymptotic approach , which is much more artificial.

It was Richard Courant, in [18], the first researcher to suggest the spaces of finite elements, in an article which failed to call much attention about. This technique was rediscovered in the 1960s, by civil engineers. It was Argyris in [4] the author of the first publication at that time, while Clough [15] was the first to have employed the expression *finite element*. Some years have passed by until part of the mathematical community took the option to theoretically examine this tool. Many of the properties then proved were already well-known and of course employed, within the application and computing framework.

As the FE – finite elements – are compared with the finite differences, it is observed that under treatment of the FE, differential operators are not **approximated**: we fully deal with their definitions on the FE approximation spaces. Making it more clear, we take the operator **restriction** to those spaces so as to solve the same equations proposed. Still, another saying: the spaces where the solutions are sought must first be approximated by spaces of FE. On these we look for the solutions that must fulfill minimization properties, just like those associated to the orthogonal projections, cf. Sect. 3.4.

The theorem discovered in 1891 by Karl Weierstrass (see pp. 23, 190, or cf. [71]) may be described as telling that the polynomials are able to very well *mimic* the set of continuous functions: no matter how precise is our measuring tool, given, on the interval $[a, b]$, an arbitrary continuous function, a polynomial may be chosen that manages to *trick* our tool. This happens because it will not be able to distinguish between the found polynomial and the function we have started with. As long as the numerical computing of a polynomial just requires algebraic operations (sum, multiplication), easily processed with a computer, this result seems to point an unbeatable track to the building of approximation algorithms.

It occurs, though, that the core of the approximation task lives not only in discovering, **finding** it, but mainly in evaluating its **numerical error**, as well as the **computational cost**, *v.g.* complexity, to be spent at implementation chores. In other words, these are competing faces of the same problem, no sense bearing to solve the former if information data are lacked on the latter ones. It can be shown that, in order to get smaller errors with the approximations assured by Weierstrass' theorem, polynomials with higher degrees are required. Besides, this approximation generates another type of errors, those linked to the digital finite precision arithmetic, the **rounding errors**.

Given $N+1$ distinct points in $[a, b]$, $\{x_j\}$, $a \leq x_0 < x_1 < \ldots < x_N \leq b$, and a function $f \in C^{N+1}(a, b)$, being $h := \max\{x_j - x_{j-1}, 1 \leq j \leq N\}$, the expression for the truncation error in the interpolation of f with the (unique) polynomial $p_N = p_N(f)$ of degree N which coincides with f on these $N+1$ points, called **interpolation nodes**, is

$$|f(x) - p_N(x)| \leq Ch^{N+1}, a \leq x \leq b, \tag{8.10}$$

where $C = C_N(f)$ is a constant associated to bounds for derivatives of f on $[a, b]$, cf. [12, p.145].

At first sight, estimate (8.10) suggests that trying to reduce the errors in polynomial approximation, with interpolation as tool, is a dead end, as it requires a larger and larger degree for these polynomials. By another look, such expression hints an alternate recipe, just the one that has inspired the construction of the method of finite elements, as hereby described.

Instead of working with global approximations, we deal with local ones, which means interpolations defined separately on subintervals of the initially considered interval. In this fashion, it is possible to have $h \to 0$ without making N to increase and, thus, to avoid the estimate (8.10) to be threatened by the value of C_N. Another way to justify this procedure is that the function f may lack the regularity level required by (8.10) for high values of N. Let us recall that piecewise polynomial functions are used in the repeated expressions for numerical integration.

We shall describe now how to approximate $H^1(\Omega)$ by quadratic finite elements, for the simpler case of $\Omega = [a, b] \subset I\!R$. Some of the technical difficulties that show up for regions Ω with dimension greater than 1 stay hidden throughout the present discussion. Nevertheless it already exhibits the most relevant aspects of the method construction. That is the aim sought when some details in the notation and even the construction are kept, despite they being justified only for larger dimensions.

Take $N > 0$ as an arbitrary integer, and consider a mesh $\mathcal{T}$ for $[a, b]$ composed of N closed intervals,[5] which we shall call **elements** and denote with $e \in \mathcal{T}$.

The extreme points and the middle point of each element are called **nodes** of this element[6] and denoted $e_1 < e_2 < e_3$; thus we have

$$\bigcup_{\{e\in\mathcal{T}\}} e = \overline{\Omega}$$

$$e \bigcap e' = \left[\begin{array}{ll} \emptyset & \text{or} \\ \{e_\imath\} = \{e'_\jmath\}, & \text{with } \{\imath, \jmath\} = \{1, 3\} \end{array}\right. .$$

Now, being

$$h_e := e_3 - e_1 = \text{ diam } e, \tag{8.11}$$

$$h := \max_{e\in\mathcal{T}}\{h_e\}, \tag{8.11'}$$

it holds

$$|e_\imath - e_\jmath| \le h, \forall e \in \mathcal{T}, \imath, \jmath = 1, 2, 3.$$

The very definition (8.11) emphasizes itself being unnecessary the choice of a **regular mesh**, *i.e.*, formed with intervals of equal length. This equal length choice even needs to be avoided in some problems, as to improve the approximation quality; see for example [34].

The space $V_h \subset H^1(\Omega)$ to be constructed is composed by continuous functions that, inside each element, coincide with a polynomial of degree ≤ 2. This way, each function $v \in V_h$ becomes completely characterized by its values in the nodes $e_\imath$, $\imath \in 1, 2, 3$, $e \in \mathcal{T}$. Being $\phi_{e_\imath} \in V_h$ the **form functions** defined for $e \in \mathcal{T}, \imath, \jmath = 1, 2, 3$ by

$$\psi_{e_\imath}(e_\jmath) := \delta_{\imath\jmath} (\delta \text{is Kronecker delta}), \tag{8.12}$$

and, by denoting

$$v_e^\imath := v(e_\imath),$$

[5] Watch up the eventual misuse of the term "partition", as there exist non void intersections for neighboring subintervals.

[6] (Watch out again, not all of them are mesh **nodes**.)

we have

$$v(x) = \sum_{\iota=1}^{3} v_e^\iota \phi_{e_\iota}(x), \forall v \in V_h, x \in e.$$

This is the so-called local representation for v, valid only at the element $e \in \mathcal{T}$. With regard to this **local basis**, v owns quite simple coefficients: they turn out to be its very values on the three nodes e_ι. On the other hand, for any element $e \in \mathcal{T}$, the functions ϕ_{e_ι} may as well be calculated at any point at the element by employing the same interpolation functions

$$\theta_\iota : [-1, 1] \to I\!R$$

given by (see the figure just below)

$$\begin{bmatrix} \theta_1(s) := s(s-1)/2 \\ \theta_2(s) := 1 - s^2 \\ \theta_3(s) := s(s+1)/2 \end{bmatrix},$$

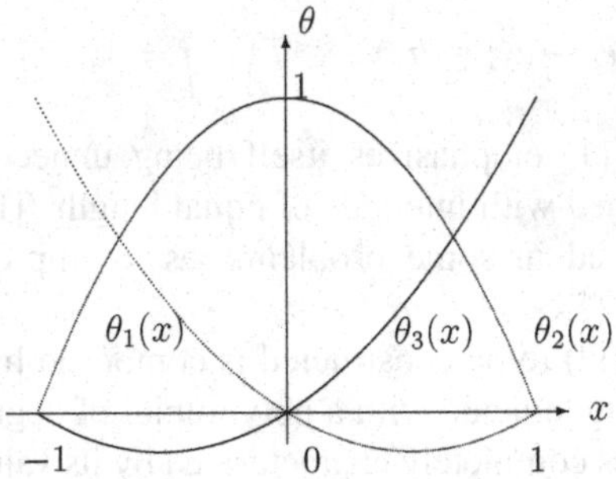

and the coordinate changes

$$\xi_e(x) := 2\frac{x - e_1}{e_3 - e_1} - 1, x \in e,$$

or else

$$x_e(\xi) := [\xi(e_3 - e_1) + e_3 + e_1]/2, -1 \le \xi \le 1.$$

In fact, we have

$$\phi_{e_\iota}(x) = \theta_\iota\left(\xi_e(x)\right), x \in e, \forall e \in \mathcal{T}.$$

The notation introduced in (8.12) is purposely ambiguous, since the same function ϕ shows up with distinct indices according being a node seen as in one or another element. Now, the family $\{\phi_{e_\iota}\}_{\iota=1,2,3;e\in\mathcal{T}}$ is a basis for $\{V_h\}$, with $N_h = 2N+1$ elements, and thus we may attach to it indices in the form $\{\phi_J\}_{1\le J\le N_h}$, which corresponds to replace the nodes indices. This amounts to have now a global indexation $\{x_J\}_{1\le J\le N_h}$ as well as a **global representation**

$$v(x) = \sum_{J=1}^{N_h} v(x_J)\phi_J(x). \tag{8.13}$$

The functions ϕ_J exhibit the profiles described in the figure below. Observe that the support of each of them intersects at most the support of other **four**. This way, the set $\{\phi_J\}$ is not more than *almost* orthogonal: the inner products $< \phi_\iota|\phi_J >$ may be distinct from zero for at most **five** pairs (ι, J). When numerically solving differential equations, this property of the finite elements assures that the linear systems generated by the resolution algorithm correspond to sparse matrices (in fact, band matrices), which lead to an important saving in computational cost.

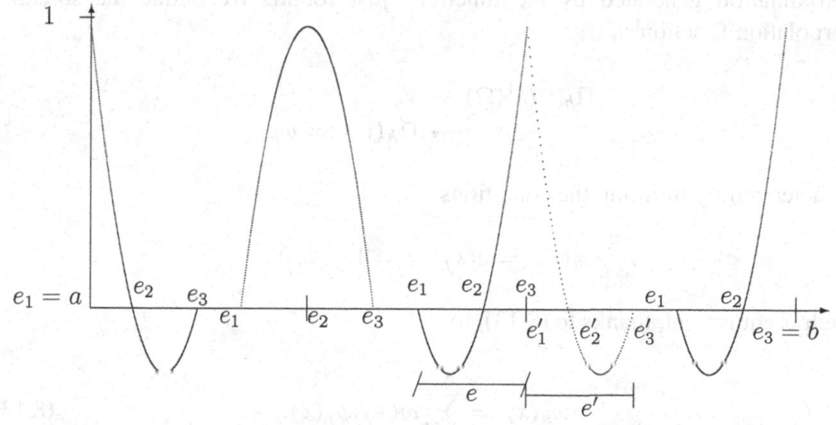

Some functions ϕ_J from the global basis we have constructed have been "tore" as to be considered in the local basis which corresponds to one of the elements where they fail to vanish. The pieces of information generated at the level of element need therefore to be "mended" in order to produce the global data. This is what is called the **assembly** of the system global matrix, from the matrices associated to the local equations.

It is a need to list now some of the properties of the finite element technique:

- Handling of the functions ϕ_j is quite simple, since essentially performed by the software on use. It is free of numerical approximation schemes, as those required when dealing with the eigenfunction expansion. These may deceive the user with their apparent advantage of generating a diagonal system, rather than just a sparse one.
- The very coefficients of a given function with respect to the basis ϕ_j already hand important data about this function – they happen to be exactly its values in the nodes e_j. This is indeed a strong property when the function to be approximated is "delivered" throughout other characteristics, for example, as the solution of a system of differential equations.
- The routine of constructing approximated solutions for a system of differential equations encompasses a huge amount of procedures also required in many other problems. This leads then to the efficient building of common program libraries.
- Indexation choice for $e_1 < e_2 < e_3$ and $x_j < x_{j+1}$ implies having the associated matrices the structure of **band matrices**. Of course this also happens to be the natural indexation in one dimension. This property ceases being valid for greater dimensions, thus making a need to search for the most convenient order for indexation. Moreover, the correspondence between global indices and local ones is not as simple, which asks for the introduction of the incidence matrices, cf. [54].

Next goal is to quantify the accuracy degree – or the error level – for the approximation generated by the functions just found. We define the so-called interpolation function

$$\begin{aligned} \Pi_h : H^1(\Omega) &\to V_h \\ w \quad &\to \Pi_h(w) := w_h, \end{aligned}$$

characterized by fulfilling the conditions

$$w_h(x_j) = v(x_j),\ j = 1, \ldots, N_k,$$

which is equivalent, thanks to (8.13), to

$$w_h(x) := \sum_{j=1}^{N_h} w(x_j)\phi_j(x). \tag{8.13'}$$

Recall the notation for h introduced in (8.11) and denote by

$$\Delta := \min_{e \in \mathcal{T}}\{h_e\},$$

to claim that, cf. [68], for every function $w \in C^2(\Omega)$, it holds

$$\|w - \Pi_h w\|_1 \leq \mathcal{C}(w)\frac{h^3}{\Delta},$$

where the constant $\mathcal{C}(w)$ depends on w – in fact, $\mathcal{C}(w) = \mathcal{C}(\|w\|_{H^2})$.

Consequently, taking into account to be the subspace $C^2(\Omega)$ dense in $H^1(\Omega)$, we state the following:

Theorem (Piecewise Quadratic Approximations) *Given $f \in H^1(\Omega)$, a sequence of real numbers $\{h_n\}$, with $h_n \to 0$, can be obtained, jointly with a sequence of functions $f^n \in V^{h_n}$ so that they fulfill*

$$\|f_n - f\|_{H^1} = \mathcal{O}(h_n^2).$$

As we have this theorem at hand, we can prove the result announced at the end of last section, which is suggested as an

Exercise The space $H^1(\Omega)$ is separable.

A more general formulation of the interpolation properties for one-dimensional finite elements with uniform mesh is the statement of the

Theorem (Piecewise Polynomial Approximations) *For given integers r, ℓ, and N, with $0 < r < \ell - 1$ and $h = 1/(N+1)$, consider the space $S_h^{\ell,r}$ of the piecewise polynomial functions of class $C^r(0,1)$ that coincide with a polynomial of degree $\leq \ell - 1$ on each subinterval. The interpolation operator on these spaces*

$$\begin{aligned} \Pi_h \colon H^1(0,1) &\to S_h^{\ell,r} \\ v &\to V_h \end{aligned}$$

fulfills

$$\|v - V_h\|_{H^1} = \mathcal{O}(h^{\ell-1}).$$

More complete formulations may be searched on [14], which also exposes – cf. pp.112–114 – details from the result that follows.

It establishes the connection between interpolation and numerical solution of differential equations with the help of the finite elements.

Let us suppose that the variational equation

$$a(u, v) = l(v), \forall v \in V, \tag{8.14}$$

fulfills Lax-Milgram conditions described on page 78. Being V_h a family of subspaces of V with the parameter $h \to 0$, on each of them we consider **discretized equation**

$$a(u_h, v_h) = l(v_h), \forall v_h \in V_h. \tag{8.15}$$

It is said that the spaces V_h generate a **convergent discretization** if, for any problem in the form (8.14) we have, for the solutions found by (8.15),

$$\lim_{h \to 0} \|u - u_h\| = 0.$$

The connection we have mentioned is assured by

Céa Lemma Under the above described conditions for $a, V, and V_h$, we can assure the existence of a constant C, which does not depend on the subspace V_h, for which

$$\|u - u_h\| \leq C \inf_{v_h \in V_h} \|u - v_h\|.$$

8.4 Compactness: Eigenvectors Bases

Consider in a Hilbert space H the operator T defined by

$$\begin{aligned} T: \quad & H \quad \to H \\ & x = \sum x^n e_n \to Tx := \sum (x^n/n) e_n, \end{aligned}$$

where $\{e_n\}$ is an orthonormal basis of H.

We claim that T is a **compact operator**.

(On this section, whenever the range of the indices for sequences or summands are omitted, it must be understood that $\jmath, k, n$ run all over $I\!N$, or some of its duly mentioned subsets, which ought to stay clear from the context.)

To check the above claim on compactness, we must take an arbitrary bounded sequence $\{x_\jmath\} \subset H$ and deduce the existence of a subsequence $\{x_{\jmath k}\}$ for which $\{Tx_{\jmath k}\}$ converges.

Step 1 (Subsequence building) Bessel inequality implies that

$$|x^n|^2 \leq \|x\|_H^2, \forall x = \sum x^n e_n \in H, \tag{8.16}$$

which assures being the real sequence $\{x_\jmath^n\}_\jmath$ bounded, no matter the (fixed) n-th component thereby considered.

Let $n = 1$. It is then possible to choose a subsequence $\{x_\jmath(1)\}$ from $\{x_\jmath\}$, such that

$$|x_\jmath^1(1) - \bar{x}^1| \to 0 \text{ if } \jmath \to \infty, \text{ for some real } \bar{x}^1.$$

By the same token, being $\{x_j^2(1)\}$ bounded, it admits a subsequence which converges, let us say, to $\bar{x}^2 \in I\!R$. Denote such a subsequence of $\{x_j(1)\}$ by $\{x_j(2)\}$, and take into account the following property that is held:

$$|x_j^2(2) - \bar{x}^2| \to 0 \text{ if } j \to \infty.$$

The present procedure may be repeated, and by this fashion we construct

$$\{x_j(1)\}, \{x_j(2)\}, \dots, \{x_j(k)\}, \dots,$$

all subsequences of $\{x_j\} \subset H$, each of them a subsequence of the previous one. Besides, they hold the property:

$$|x_j^k(k) - \bar{x}^k| \to 0 \text{ if } j \to \infty.$$

The convergent subsequence whose existence we have claimed is reached by the so-called Cantor diagonal process (already employed in Sect. 5.2).
Let us denote it by $\{\tilde{x}_\imath\}$, being it defined by

$$\tilde{x}_\imath := x_\imath(\imath),\ \imath \in I\!N.$$

Step 2 (Convergence checking) We claim that the vector

$$\bar{y} := \sum (\bar{x}^n/n) e_n \in H$$

is the limit of the sequence $\{T\tilde{x}_\imath\}$, that is

$$\lim_{\imath\to\infty} \|T\tilde{x}_\imath - \bar{y}\|^2 = \lim_{\imath} \sum_n |\tilde{x}_\imath^n - \bar{x}^n|^2/n^2 = 0. \tag{8.17}$$

We observe at once that $\bar{y} \in H$ because (cf. (8.16))

$$\sum |\bar{x}^n/n|^2 \le K \sum n^{-2},$$

being K an upper bound for $\|x_j\|^2$.

Take now $\epsilon > 0$ arbitrary. The convergence of the series $\sum n^{-2}$ implies the possibility of determining $N = N(\epsilon)$ such that

$$\sum_{n=N}^{\infty} n^{-2} < \epsilon/4K$$

and, as a consequence,

$$\sum_{n=N}^{\infty} |\tilde{x}_\iota^n - \bar{x}^n|^2/n^2 < \epsilon/2. \tag{8.18}$$

It suffices now, in order to reach (8.17), to verify that

$$\sum_{n=1}^{N-1} |\tilde{x}_\iota^n - \bar{x}^n|^2/n^2 < \epsilon/2.$$

Well, since all $(N-1)$ components $\tilde{x}_\iota^n$ converge (to $\bar{x}^n$, respectively) if $\iota \to \infty$, then it is possible to find $I = I(\epsilon)$ for which

$$|\tilde{x}_s^n - \bar{x}^n|^2/n^2 < \epsilon/2^N \left[\begin{array}{l} s \geq I \\ n = 1, \ldots, N-1 \end{array} \right. . \tag{8.19}$$

By combining (8.18) to (8.21) we reach, for $\iota > I$, the inequality

$$\|T\tilde{x}_\iota - \bar{y}\|^2 < \epsilon.$$

Exercise 8.3 Observe that with this same proof scheme, we can deduce the compactness of the so-called Hilbert cube, the subset

$$\mathcal{K} := \{x = (x^n) \in \ell^2; |x^n| \leq 1/n\}.$$

The operator T above is more than a (maybe) hard example of a compact operator: it explains how the members of a large compact operators family act. Such is the family that hosts the compact operators as well as the so-called self-adjoint ones, which means: they fulfill the condition

$$(Tx|y) = (x|Ty), \forall x, y \in V. \tag{8.20}$$

These latter operators own an important characteristic: their images $T(V)$ exhibit an expansion with respect to an orthonormal basis composed by eigenvectors.

We list below the main properties associated to their spectra.

a) All eigenvalues from T are real, and they form an at most countable set $\{\lambda_\iota; \iota \in \mathbb{N}\}$ whose only limit point is zero, $\lambda_\iota \to 0$ if $\iota \to \infty$.
b) To each eigenvalue λ_ι, there corresponds a subspace V_ι, of finite dimension δ_ι, composed by the set of the eigenvectors associated to λ_ι (joined by the null vector, of course).
c) Eigenvectors that correspond to distinct eigenvalues are orthogonal to each other.

d) Each eigenvalue holds the description from Courant **mini-max principle:**[7]

$$\begin{bmatrix} |\lambda_1|^2 = \sup\{(Tx|Tx);\ \|x\| = 1, x \in V\} \\ |\lambda_\imath|^2 = \sup\{(Tx|Tx);\ \|x\| = 1, x \in V_\jmath^\perp, \imath > \jmath \geq 1\} \end{bmatrix}. \tag{8.21}$$

8.5 Unbounded Operators

A natural conclusion about unbounded operators, from a reader who has been introduced to the Functional Analysis aisles only through the present text, is that they are bound to the role of unwanted invitees. Moreover, inspired by real function grounds, continuity at no point – cf. Theorem 2.1 from Sect. 2.11 – is an anomaly prone to be rejected by our day-to-day function regularity searches. Thus, these operators existence would only be justified to give a hand to creation of counterexamples.

Being this the last section on our last chapter, it is worth to bring some words that could be an aid to erase such a parti-pris.[8] For that, the first smart step is to invite a well-accepted operator sample.

Provisionally take the Laplacian differential operator $-\Delta : C_0^\infty \to C_0^\infty$ on the whole of $I\!R^n$. This functional framework fails to be handy as regards to the domain choice, having in mind that solutions to be searched commonly lack such a strong smoothness. Another temptation concerns the unmentioned norm choice. Inspired by $-\Delta : H^2(I\!R^n) \to L^2(I\!R^n)$, we will be dealing with the comfortable bounded operator borough. But, in several contexts, it is worth to deal with the same environment for the departure and arrival spaces, so it would be better to keep the same norm for domain and range. There is a toll to be paid, though, as seen in the development that follows.

From now on we deal with linear operators T whose domain $D(T)$ is a proper subspace of the main space under use, say $(N_1, ||\cdot||_1)$, while its range is a subspace of $(N_2, ||\cdot||_2)$. The considered operators are not necessarily bounded, that's why we call them **unbounded**. We also must consider on $D(T)$ the so-called graph norm: it assigns to each $v \in D(T)$ the value of its norm as an element in the graph of T, under the associated product space norm. Let's tell it in a shorter way: $||v||_T := ||v||_1 + ||Tv||_2$.

We refer to the operator T as being **closed** whenever its graph is a closed subset of $N_1 \times N_2$, under the product norm. It may be proven the equivalence of this definition and the one which requires being $D(T)$ a Banach space under the graph norm.

Now, an application:

---

[7] Cf. [17], or Chap. 6 in [19].

[8] Plenty of up-to-date information for this topic, very clearly written by C. Cheverry and N. Raymond, is found in [13].

Claim 1 Take $N_1 = N_2 = L^2(I\!R^n)$ and the operator $-\Delta$ with $H^2(I\!R^n)$ as its domain. This is a closed operator.

An operator is said to be **closable** if one of its extensions is a closed operator. As long as such a definition is posed, one naturally deduces that not all operators are closable. This is true (see Exercise 2.21, [13]). And what about all closed extensions for T? Is there one that mimics this operator in a better way than all other ones? Indeed it can be found, being given by the closure of the graph of T, as always under the product norm. And as a gift – or a mnemonics – it gets the same name, the **closure of the operator** T. We show an example of these ideas, in part linking back to this section start, with the

Claim 2 The closure of the minus Laplacian operator $-\Delta$ when taken $C_0^\infty(I\!R^n)$ as its domain, as a subspace of $L^2(I\!R^n)$, is given by, again, the action of minus Laplacian but now on $H^2(I\!R^n)$.

Appendix A
Recent References

Our text [25], which gave birth to the present one, came out of press more than two decades ago. Mandatory, thus, to look at references published since then – that is what this appendix aims to. Some of the publications quoted herein are even from earlier years, later editions that have lately appeared endorse their importance on our discussion area.

During this time period, the road taken by scientific computing was already paved by parallel/distributed programming and the consequent help calling to graph theory. It has also shown this research environment discovery by different trends, between them a huge presence of biology, medicine, ecology, and economy. These areas have all switched from computational humble demands to very heavy requirements. In the problems their models deal with, nondeterministic assumptions show up as a solid need. As a consequence, the steady presence of stochastic results have become a strong asset. These two topics – graphs and randomness analyses – being scarce in our list, deserve to be emphasized.

Let's start with [53], an already classical piece, which bears plenty of information for anyone leaning to surf on the finite-element waves. We also find [63] as worth reminding, thanks to its precise theoretical approach to some mathematical models for real phenomena, jointly with [3], which strolls through applications supported by concepts it thoroughly exposes.

All books in the list which follows partially share the intentions and contents of the present text. The first also deals with shape optimization, while the second one discusses tensors and wavelets, carrying out an informal exposition: [10, 44, 52, 56, 64, 66]. We should also point out that the just refereed book [56] browses through signal theory and its Fourier treatment, which it starts by discussing DFT – the Discrete Fourier Transform – while the previously listed [52] has dealt with quite a bunch of results on semigroups.

C. A. de Moura, *Functional Analysis Tools for Practical Use in Sciences and Engineering*, https://doi.org/10.1007/978-3-031-10598-2

Besides the traditional excuses presented by authors for choices made on lists preparation, mine is both short and bears a disconnected appearance: let me finish it with [43, 59] as the two ending references. While appraisal for the first goes to its choice of the teaching methodology that makes the road from examples to theoretical basis, the other one presents an impressive large amount of useful examples.

References

1. Adams, R.A.: Sobolev Spaces. Academic Press, New York (1975)
2. Agmon, S.: Lectures on Elliptic Boundary Value Problems. Van Nostrand, Princeton (1965)
3. Alt, H.W.: Linear Functional Analysis—An Application-Oriented Introduction (translation from the 6th German edition). Universitext, Springer, London (2016)
4. Argyris, J.H.: Energy theorems and structural analysis, part i: General theory. Aircraft Engin. **26–27**, 347–356, 383–387, 397 (1954)
5. Aubin, J.P.: Approximation of Elliptic Boundary Value Problems. Wiley, New York (1972)
6. Bachman, G., Narici, L.: Functional Analysis. Academic Press, New York (1966)
7. Bartle, R.G.: The Elements of Integration. Wiley, New York (1966)
8. Beltrami, E.J., Wohlers, M.R.: Distributions and the Boundary Values of Analytic Functions. Academic Press, New York (1966)
9. Boas Jr., R.P.: A Primer of Real Functions, Carus Monograph 13. Mathematical Association of America, New York (1960).
10. Botelho, F.S.: Functional Analysis, Calculus of Variations and Numerical Methods for Models in Physics and Engineering. CRC Press, Boca Raton, FL (2021)
11. Burgarelli, D., Kischinhevsky, M., Biezuner, R.J.: A new adaptive mesh refinement strategy for numerically solving evolution PDE's. J. C. Appl. Math. **196**, 115–131 (2006)
12. Campos filho, F.F.: Algoritmos Numéricos, 2a. ed. LTC, Rio de Janeiro (2007)
13. Cheverry, C., Raymond, N.: Handbook of Spectral Theory. https://hal.archives-ouvertes.fr/cel-01587623v3, France (2019)
14. Ciarlet, P.G., Lions, J.-L. (eds.): Handbook of Numerical Analysis, vol. II (Part 1). North-Holland, Amsterdam (1991)
15. Clough, P.W.: The finite element method in plane stress analysis. In: Proceedings of the 2nd ASCE Conference on Electronic Computation, Pittsburgh, PA, USA (1960)
16. Collatz, L.: Functional Analysis and Numerical Mathematics. Academic Press, New York (1966)
17. Courant, R.: Über die Eigenwerte bei den Differenzengleichungen der mathematisches Physik. Mathematische Zeitschrift **7**, 1–57 (1920)
18. Courant, R.: Variational methods for the solution of problems of equilibrium and vibration. Bull. Am. Math. Soc. **49**, 1–23 (1943)
19. Courant, R., Hilbert, D.: Methods of Mathematical Physics. Interscience, New York (1953)
20. Courant, R., Friedrichs, K.O., Lewy, H.: Über die partiellen Differenzengleichungen der mathematisches Physik. Math. Ann. **100**, 32–74 (1928)
21. Davis, M.: A First Course in Functional Analysis. Gordon and Breach, New York (1966)

C. A. de Moura, *Functional Analysis Tools for Practical Use in Sciences and Engineering*, https://doi.org/10.1007/978-3-031-10598-2

22. de Moura, C.A.: Análise Funcional e Aplicações, I Escola de Matemática Aplicada, LAC-CBPF/CNPq, Rio de Janeiro (1978)
23. de Moura, C.A.: Análise Funcional—um Roteiro. In: Elementos Finitos e Aplicações à Mecânica dos Fluidos, pp. i–vi, 1–180. V Escola de Matemática Aplicada, LNCC/CNPq, Rio de Janeiro (1984)
24. de Moura, C.A.: DES: an Explicit, Really Quadratic 2-level Scheme for the Diffusion Equation. J. Comp. Inf. **3**(1), 99–115 (1993)
25. de Moura, C.A.: Análise Funcional e Aplicações—Posologia. Editora Ciência Moderna, Rio de Janeiro (2002)
26. de Moura, C.A., Burgarelli, D.: Tópicos de Análise Funcional na Computação Científica—São Carlos, SP: SBMAC (Notas Mat. Aplicada 45) (2012)
27. de Moura, C.A., Kubrusly, C.S. (eds.): The Courant–Friedrichs–Lewy (CFL) Condition. Birkhäuser, Basel (2013)
28. Dennis Jr., J.E., Schnabel, R.B.: Numerical methods for unconstrained optimization and nonlinear equations. Prentice-Hall, Englewood Cliffs (1983)
29. Dirac, P.A.M.: The Principles of Quantum Mechanics, 4th ed. Oxford University Press, London (1957)
30. Fano, G.: Metodi Matematici della Mecanica Quantistica. Zanichelli, Bologna (1967)
31. Fleming, W.H.: Functions of Several Variables. Addison-Wesley, Reading (1965)
32. Friedman, A.: Partial Differential Equations. Holt, New York (1969)
33. Gel'fand, I.M., Shilov, G.E.: Generalized Functions, vol. I–IV. Academic Press, New York (1964)
34. George, P.L.: MODULEF: Génération automatique de maillages. Collection Didactique, INRIA, Rocquencourt, France (1986)
35. Gilioli, A.: Equações Diferenciais Parciais Elíticas. IMPA, Rio de Janeiro (1975)
36. Goldstein, J.A.: Semigroups of linear operators and applications. Oxford University Press, London (1985)
37. Graves, L.M.: The Theory of Functions of Real Variables, 2nd ed. McGraw-Hill Book, New York (1956)
38. Hellwig, G.: Partial Differential Equations. Blaisdell, New York (1964)
39. Hönig, C.S.: Análise Funcional e Aplicações. EdUSP, São Paulo (1970)
40. Iório Jr., R., Iório, V.M.: Equações Diferenciais Parciais: uma Introdução, Projeto Euclides. IMPA, Rio de Janeiro (1988)
41. John, F.: Partial Differential Equations. Springer, Berlin (1974)
42. Kolmogorov, A., Fomin, S.: Elementos de la Teoría de Funciones y del Análisis Funcional (ed.) MIR, Moscow (1972)
43. Krishnan, V.K.: Textbook of Functional Analysis—A problem-oriented approach, 2nd ed. PHI Learning, New Delhi (2014)
44. Kumaresan, S., Sukumar, D.: Functional Analysis—a first course. Alpha Science International, Oxford (2021)
45. Lax, P.D.: On the existence of Greens's function. Proc. Am. Math. Soc. **3**, 526–531 (1952)
46. Lax, P.D.: Real Variables, NYU Lecture Notes. Courant Institute, New York (1970)
47. Lax, P.D.: Functional Analysis, Pure and Applied Mathematics: A Wiley Series of Texts. Wiley, Hoboken (2002)
48. Lax, P.D., Richtmyer, R.D.: Survey of stability of the linear finite difference schemes. Commun. Pure Appl. Math. **9**, 267–293 (1956)
49. Mackey, G.W.: Mathematical foundations of quantum mechanics. In: Mathematical Physics Monograph Series. W.A. Benjamin, Inc., New York (1963)
50. Medeiros, L.A., Milla-Miranda, M.: Espaços de Sobolev—Introdução aos Problemas Elíticos não Homogêneos. IM/UFRJ, Rio de Janeiro (2000)
51. Mizohata, S.: The Theory of Partial Differential Equations. Cambridge Press, Cambridge (1973)
52. Moroşanu, G.: Functional Analysis for the Applied Sciences. Universitext, Springer, London (2019)

53. Oden, J.T., Demkowicz, L.F.: Applied Functional Analysis, 3rd edn. Textbooks in Mathematics. CRC Press, Boca Raton (2018)
54. Oden, J.T., Reddy, J.N.: An Introduction to the Mathematical Theory of Finite Elements. Wiley Interscience, New York (1976)
55. Pombo Jr., D.P.: Introdução à Análise Funcional. EdUFF, Niterói, Brazil (1999)
56. Provenzi, E.: From Euclidean to Hilbert spaces—Introduction to Functional Analysis and its applications, Mathematics and Statistics Series. Wiley, Hoboken (2021)
57. Richtmyer, R.D., Morton, K.W.: Difference Methods for Initial-value Problems, 2nd edn. Interscience Publishers, New York (1967)
58. Riesz, F., Nagy, B.: Functional Analysis. Ungar, New York (1966)
59. Robinson, J C.: An introduction to Functional Analysis. Cambridge University, Cambridge (2020)
60. Royden, H.L.: Real Analysis. MacMillan, London (1968)
61. Rudin, W.: Principles of Mathematical Analysis. McGraw Hill, New York (1963)
62. Rudin, W.: Real and Complex Analysis. McGraw Hill, New York (1966)
63. Sacks, P.: Techniques of Functional Analysis for Differential and Integral Equations. In: Mathematics in Science and Engineering. Elsevier/Academic Press, Amsterdam (2017)
64. Sasane, A.: A friendly approach to Functional Analysis. In: Essential Textbooks in Mathematics. World Scientific, Hackensack (2017)
65. Schwartz, L.: Méthodes Mathématiques pour les Sciences Physiques. 2ème éd. Hermann, Paris (1965)
66. Shima, H.: Functional Analysis for Physics and Engineering —an introduction. CRC Press, Boca Raton (2016)
67. Simmons, G.F.: Introduction to Topology and Modern Analysis. McGraw Hill, New York (1963)
68. Strang, G., Fix, G.J.: An Analysis of the Finite Element Method. Prentice-Hall, Englewood Cliffs (1973)
69. Taylor, A.: Introduction to Functional Analysis. Wiley, New York (1958)
70. Titchmarsh, E.C.: The Theory of Functions, 2nd ed. Oxford University Press, London (1939)
71. Weierstrass, K.: Mathematische Werke. Mayer und Müller, Berlin (1894)
72. Yosida, K.: Functional Analysis. Springer, Berlin (1971)

Index

C. A. de Moura, *Functional Analysis Tools for Practical Use in Sciences and Engineering*, https://doi.org/10.1007/978-3-031-10598-2

O

P

Q

R

S

Printed in the United States
by Baker & Taylor Publisher Services